Sasmita Panda
Gagan Kumar Panigrahi
Surendra nath Padhi

Wild Animals Of India

Anchor Academic
Publishing

Panda, Sasmita, Panigrahi, Gagan Kumar, Padhi, Surendra nath: Wild Animals Of India, Hamburg, Anchor Academic Publishing 2016

Buch-ISBN: 978-3-96067-014-8
PDF-eBook-ISBN: 978-3-96067-514-3
Druck/Herstellung: Anchor Academic Publishing, Hamburg, 2016
Covermotiv: © pixabay.com

Bibliografische Information der Deutschen Nationalbibliothek:
Die Deutsche Nationalbibliothek verzeichnet diese Publikation in der Deutschen Nationalbibliografie; detaillierte bibliografische Daten sind im Internet über http://dnb.d-nb.de abrufbar.

Bibliographical Information of the German National Library:
The German National Library lists this publication in the German National Bibliography. Detailed bibliographic data can be found at: http://dnb.d-nb.de

All rights reserved. This publication may not be reproduced, stored in a retrieval system or transmitted, in any form or by any means, electronic, mechanical, photocopying, recording or otherwise, without the prior permission of the publishers.

Das Werk einschließlich aller seiner Teile ist urheberrechtlich geschützt. Jede Verwertung außerhalb der Grenzen des Urheberrechtsgesetzes ist ohne Zustimmung des Verlages unzulässig und strafbar. Dies gilt insbesondere für Vervielfältigungen, Übersetzungen, Mikroverfilmungen und die Einspeicherung und Bearbeitung in elektronischen Systemen.

Die Wiedergabe von Gebrauchsnamen, Handelsnamen, Warenbezeichnungen usw. in diesem Werk berechtigt auch ohne besondere Kennzeichnung nicht zu der Annahme, dass solche Namen im Sinne der Warenzeichen- und Markenschutz-Gesetzgebung als frei zu betrachten wären und daher von jedermann benutzt werden dürften.

Die Informationen in diesem Werk wurden mit Sorgfalt erarbeitet. Dennoch können Fehler nicht vollständig ausgeschlossen werden und die Diplomica Verlag GmbH, die Autoren oder Übersetzer übernehmen keine juristische Verantwortung oder irgendeine Haftung für evtl. verbliebene fehlerhafte Angaben und deren Folgen.

Alle Rechte vorbehalten

© Anchor Academic Publishing, Imprint der Diplomica Verlag GmbH
Hermannstal 119k, 22119 Hamburg
http://www.diplomica-verlag.de, Hamburg 2016
Printed in Germany

PREFACE

This book on "Wild Animals Of India" has been written with a motive to provide information at a glance to the readers interested in wildlife. The animals chosen for study were of greater interest because they have been declared endangered, critically endangered, vulnerable or priority species in the IUCN Red data book and most of the species are included as course material at undergraduate and post graduate levels in Indian Universities. Hence this is an endeavour to create awareness among student community on Wild Life Biology.

Efforts have been made to acquaint the readers with geographical distribution, habit, habitat, reproductive behavior and conservation measures of the animals. Help from different sources have been taken to collect information written in this book. We gratefully acknowledge all the sources.

We are indebted to our parents, teachers, friends and well wishers for their encouragement. The authors are grateful to Prof. Dr. A.K. Panda, former principal of Ravenshaw College, Cuttack and Prof. Dr. B.K. Patnaik, for their guidance. Dr. R.K. Mahapatra, Smt. Sunita Sarangi, Principal, Jatni College, Jatni deserve our gratitude for their technical assistance and encouragement.

We welcome suggestions from the readers for improvement of the book.

Authors

Sasmita Panda

G.K. Panigrahi

S.N. Padhi

Dedicated to

Our Parents and well wishers

Our teachers and students

"Mini" – Who is no more to share the pleasure of this publication.

TABLE OF CONTENTS

PREFACE 1

CHAPTER 1: INDIAN ELEPHANT 5

CHAPTER 2: TIGER 8

CHAPTER 3 CHEETAH 16

CHAPTER 4; LION 20

CHAPTER 5: NILGAI 26

CHAPTER 6: BISON 28

CHAPTER 7: ANTELOPE 32

CHAPTER 8: INDIAN RHINOCEROS 36

CHAPTER 9: IRRAWADDY DOLPHIN (SNUBFIN DOLPHIN) 41

CHAPTER 10: INDIAN PEAFOWL 46

CHAPTER 11: GHARIAL 49

CHAPTER 12: SEA TURTLES 53

REFERENCES 60

CHAPTER 1
INDIAN ELEPHANT

CLASSIFICATION
Kingdom: Animalia, Phylum: Chordata, Class: Mammalia, Order: Proboscidea, Family: Elephantidae, Genus: *Elephas*, Species: *maximus*, Sub-Species: *indicus*

GENERAL CHARACTERS

The **Indian elephant** (*Elephas maximus indicus*) is one of three recognized sub species of the Asian elephant and native to mainland Asia, India, Nepal, Bangladesh, Bhutan, Myanmar, Thailand, Malay Peninsular, Laos, China, Cambodia, and Vietnam. Since 1986, *Elephas maximus* has been listed **as endangered by IUCN**. Asian elephants are threatened by habitat loss, degradation and fragmentation. In general, Asian elephants are smaller than African elephants and have the highest body point on the head. The tip of their trunk has one finger-like process. Their back is convex or level. Indian elephants reach a shoulder height of between 2 and 3.5 m (6.6 and 11.5 ft), weigh between 2,000 and 5,000 kg (4,400 and 11,000 lb), and have 19 pairs of ribs. Their skin color is lighter than that of *maximus* with smaller patches of de pigmentation, but darker than that of *sumatranus*. Females are usually smaller than males, and have short or no tusks.

Indian elephants have smaller ears, but relatively broader skulls and larger trunks than African elephants. Toes are large and broad. Unlike their African cousins, their abdomen is proportionate with their body weight but the African elephant has a large abdomen as compared to the skulls.

HABIT AND HABITAT

The elephants are regarded as megaherbivores feeding on grasses, plants and trees. They eat between 149 and 169 kg (330-375 lb.) of vegetation daily. Sixteen to eighteen hours, or nearly 80% of an elephant's day is spent on feeding. They consume grasses, small plants, bushes, fruit, twigs, tree bark, and roots. Tree bark is a favorite food source for elephants. It contains calcium and roughage, which aids in digestion. Tusks are used to carve into the trunk and tear off strips of bark. They require about 68.4 to 98.8 L (18 to 26 gal.) of water daily, but may consume up to 152 L (40 gal.). An adult male elephant can drink up to 212 L (55 gal.) of water in less than five minutes. To supplement the diet, elephants dig up earth to obtain salt and minerals. The tusks are used to churn the ground. The elephant then places dislodged pieces of soil into its mouth, to obtain nutrients. Frequently these areas result in holes that are several feet deep and vital minerals are made accessible to other animals. Over time, African elephants have hollowed out deep caverns in a volcano mountainside on the Ugandan border, to obtain salt licks and minerals. Hills have been carved by Asian elephants in India and Sumatra searching for salt and minerals. These carved areas in the landscape provide valuable food and shelter resources for a diverse array of native wildlife.

REPRODUCTION

There is no specific mating season for the Indian Elephants. Both males and females become sexually mature at about 14 years of age. The gestation period is usually 20- 22 months and the females have the capability of giving birth to a calf in every four to five years. At birth elephants can be one meter tall and weigh around ninety kilograms. They can stand soon after the birth. Female elephants supervise their young ones for several years after weaning. Females may have up to 12 calves in their lifetime of about 70 years. As the time for giving birth approaches, the female will seek close contact with another female in her family unit for protection during labor. Sometimes the entire family unit circles around a female giving birth, protecting her from all sides. Females give birth while standing. The birth itself lasts only a few minutes. A single calf is usually born head and forelegs first. Twins have been documented, but are extremely rare. Mothers will consume the afterbirth to avoid detection by predators. On an average, newborn calves stand about 1 m (3 ft.) high and weigh 120 kg (264 lb.) at birth. Newborn male African elephants may weigh up to 165 kg (364 lb.). Newborn Asian elephant calves weigh about 91 kg (200 lb.). Calves are able to stand on their own within minutes of birth. The mother and other females help guide the calf to nurse almost immediately. The trunk of the calf is still short, so it uses its mouth to nurse. Calves are able to walk within one to two hours of birth. Within two days, calves are strong enough to join the rest of the herd, that is waiting patiently nearby.

DEVELOPMENT

Mothers, aunts, sisters, and the matriarch are very important to calf development. The pace of the herd is adjusted, so the young can keep up. Calves learn which plants are edible and ways to acquire them, by watching their elders. Mothers and aunts are in almost constant affectionate contact with the young, offering guidance and assistance. Calves nurse for the first six months of life. Elephant milk is high in fat and protein (100 times more than the protein contained in cow's milk). On average, calves drink about 10 L (21 pt.) a day. Calves begin to experiment with their developing trunks between four and six months of age by picking grasses and leaves to supplement their diet. Weaning from milk gradually follows this process. Calves are not completely weaned until they are over two years of age and may weigh between 850-900 kg (1,874-1,984 lb.).

CONSERVATION STATUS

The population of the Indian elephants lie somewhere between 38,000 and 51,000. They have been continuously haunted by the people for the food, for ivory and for the domestic stock. They have suffered the great habitat loss due to the deforestation and agriculture.

CHAPTER 2
TIGER

CLASSIFICATION
Kingdom: **Animalia**, Phylum: **Chordata**, Class: **Mammalia**, Order: **Carnivora**, Family: **Felidae**, Genus: ***Panthera***, Species: ***tigris***

GENERAL CHARACTERS
The tiger is the national animal of India, Bangladesh, Vietnam, Malaysia and South Korea. The tiger (*Panthera tigris*) is the largest cat species, reaching a total body length of up to 3.38 m (11.1 ft) over curves and weighing up to 388.7 kg (857 lb) in the wild. Its most recognisable feature is a pattern of dark vertical stripes on reddish-orange fur with a lighter underside. The species is classified in the genus *Panthera* with the lion, leopard, jaguar and snow leopard. Tigers are apex predators, primarily preying on ungulates such as deer and bovids. They are territorial and generally solitary but social animals, often requiring large contiguous areas of habitat that support their prey requirements. This, coupled with the fact that they are indigenous to some of the more densely populated places on Earth, has caused significant conflicts with humans. They range from the Siberian taiga to open grasslands and tropical mangrove swamps. They have been classified as **endangered by IUCN**. The global population in the wild is estimated to number between 3,062 and 3,948 individuals, down from around 100,000 at the start of the 20th century, with most remaining populations occurring in small pockets isolated from each other, of which about 2,000 exist on the Indian subcontinent.Major reasons for population decline include habitat destruction, habitat fragmentation and poaching. The extent of area occupied by tigers is estimated at less than 1,184,911 km^2 (457,497 sq mi), a 41% decline from the area estimated in the mid-1990s.They may live up to 26 years.

HABIT AND HABITAT

Tigers eat a variety of prey ranging in size from termites to elephant calves. However, an integral component of their diet are large-bodied prey weighing about 20 kg (45 lb) or larger such as moose, deer species, pigs, cows, horses, buffalos and goats. Occasionally they may consume tapirs, elephant and rhinoceros calves, bear species, leopards and Asiatic wild dogs. Tigers live in a diverse array of habitats such as tropical rainforests, mangrove swamps, evergreen forests, grasslands, savannahs, and rocky areas. Tigers mainly rely on their sense of sight and hearing rather than on smell when hunting prey. They cautiously stalk their prey from the rear in attempt to get as close as possible to their unsuspecting prey. Then they attempt to take down their prey with a powerful bite to the neck and/ or throat. They may consume up to 40 kg (88 pounds) of meat at one time. It is estimated that every tiger consumes about 50 deer-sized animals each year, about one per week.

SPECIAL CHARACTERS

Tigers have **forward facing eyes** rather than one on each side of their head. This provides binocular vision because each eye's field of vision overlaps creating a three dimensional image. Binocular vision enables them to accurately assess distances and depth which is extremely useful for maneuvering within their complex environment and stalking prey. They have more rods (responsible for visual acuity for shapes) in their eyes than cones (responsible for color vision) to assist with their night vision. The increased number of rods allows them to detect movement of prey in darkness where color vision would not be useful. They have a structure at the back of the eye behind the retina called the tapetumlucidum that enables them to have better night vision. This mirrorlike structure reflects light (that has not already been absorbed by the eye) back into the eye a second time to help produce a brighter image. The tapetumlucidum causes their eyes to glow at night when a light is shone on them. The eyes have large lenses and pupils that increase the amount of light let into the eye. This characteristic helps the tiger with night vision and when there are low light levels available. Research suggests that cats in general are capable of seeing the colors green, blue and possibly red, just in less saturation or strength than we see them. In addition to the upper and lower eyelids that protect the eye, cats and other animals such as crocodilians (alligators, crocodiles, etc.) have a nictitating membrane on each eye that helps keep it moist and removes dust from the surface. In general cats require only about 1/6 the light humans do to see.

Tigers have a well-developed **sense of touch** that they use to navigate in darkness, detect danger and attack prey. They have five different types of whiskers that detect sensory information and are differentiated by their location on the body. Whiskers differ from guard hairs in that they are thicker, more deeply rooted in the skin and surrounded by a small capsule of blood. The root of the whisker displaces the blood when the whisker comes into contact with something thereby amplifying the movement. Sensory nerves detect this movement and send signals to the brain for interpretation. The mystacial whiskers are located on the tiger's muzzle (snout) and are used when attacking prey and navigating in the dark. The tiger uses these whiskers to sense where they should inflict a bite. When navigating through darkness the tiger's pupils dilate to let more light enter the eye to increase their vision. The dilated pupils of their eyes assist their night vision but makes focusing on objects up-close difficult. The tiger's mystacial whiskers help it feel its way through the dark. Superciliary whiskers are located above the eyes. Cheek whiskers are located just behind the mystacial whiskers on the cheeks. Carpal whiskers are located on the back of the tiger's front legs. Tylotrich whiskers are located randomly throughout the body. The facial whiskers are about 15 centimeters (six inches) in length. The facial area has numerous sensory neurons that can detect even the slightest change in air pressure when passing by an object.

The tiger's **sense of hearing** is the most acute all its senses and is mainly used for hunting. Their ears are capable of rotating, similar to a radar dish, to detect the origins of various sounds such as the high-frequency sounds produced by prey in the dense forest undergrowth. Cats in general are more sensitive to high-pitched sounds than humans are. Cats may hear sounds up to 60 kHz whereas a human's upper auditory range is about 20 kHz. This sensitivity enables them to detect the high-pitched sounds emitted by prey and their movements.

The tiger's **sense of smell** is not as acute as some of its other senses and is generally not used for hunting. They have small amounts of odor-detecting cells in their nose and a reduced olfactory region in the brain that identifies various scents. It mainly uses its sense of smell for communicating information with one another such as territories and reproductive status. Tigers, like other carnivores, have a Jacobson organ in the roof of their mouth. The Jacobson organ is a pouch-like structure located directly behind the front incisors. It has two small openings that direct scent particles from the air as the tiger inhales to nerves located within the structure. The nerves then transmit the message to the olfactory region in the brain that identifies the scent. They exhibit a behavior called flehman, in which they pick up a scent on their upper lip and curl it upwards towards their nose to detect scents. This behavior makes the tiger appear to be snarling but without any sound. Tigers seem to be able to **taste** salt, bitter and acidic flavors and to a lesser degree sweetness. Cats in general possess only about 500 taste buds compared to a human's 9,000. Therefore taste buds are speculated to have a minimal role in their survivability.

ADAPTATIONS

The tiger's **striped coat** helps them blend in well with the sunlight filtering through the tree tops to the jungle floor. The tiger's seamless camouflage to their surroundings is enhanced because the striping also helps break up their body shape, making them difficult to detect for unsuspecting prey. The sense of hearing is so sharp that they are capable of hearing infrasound, which are sound waves below the range of normally audible sound (20 hertz). Tigers use infrasound to communicate over long distances or dense forest vegetation because the sound is capable of passing through a variety of mediums such as trees and mountains.

Tigers utilize a variety of **vocalizations** to communicate over long distances. Roaring is produced in a variety of situations such as taking down large prey, signaling sexual receptivity and females calling to their young. These roars may be heard from distances over 3 km (1.8 mi.).Moaning vocalizations are described as a subdued roar made while tigers are calmly walking with their heads in a downward position. This vocalization is audible for distances less than 400 m (440 yd.).Chuffing are friendly vocalizations that generally consist of a soft brrr sound. These vocalizations are primarily used for greetings between tigers and only audible at close range.

Tigers use their **tails to communicate with one another**. A tiger is relaxed if their tail is loosely hanging. Aggression is displayed by rapidly moving the tail from side to side or by holding it low with occasional intense twitches. They may enhance their olfactory communication by using visual markings such as scrapes on the ground and trees.

Adult males and females both **communicate** to one another by marking their territories. An adult tiger will usually define the boundary of its territory by **spraying urine** because of the strong odor associated with it can last up to 40 days but they may also **use faeces** for marking. All cats, including tigers, have a distinct scent associated with them due to their individualized scent glands. The individualized scent helps cubs track their mother's path and serves to identify particular individuals. Cats have scent glands between their toes, tail, anus, head, chin, lips, cheeks, and facial whiskers. A common behavior of the domestic cat is rubbing against its caregiver's legs, face and furniture. The cat's intention is to leave its individualized scent to communicate with other animals its territory and belongings. The scientific community is currently trying to train dogs to detect some individualized tiger scents to assist with the estimation of wild tiger populations.The scent glands between a tiger's toes leave behind an individualized scent that enables a cub to follow its mother's footsteps.

BEHAVIOUR

Tigers are territorial and usually **solitary** in nature. Their social system is connected through visual signals, scent marks and vocalizations. Tigers are usually solitary in nature, interacting briefly only for mating purposes and occasionally to share their kill. However, there has been a few rare instances documented in which tigers have collaborated on a hunt, similar to a pride of lions. The size of tiger territories varies greatly by locality, season and prey density (the amount of prey in a given area). In areas with high prey densities, tiger territories tend to be smaller in size because ample prey may be found in smaller vicinity. For male tigers in Ranthambhore India; the prey concentrations are high and male tigers have territories that range in size from 5 to 150 km^2 (2 to 60 km^2). In Siberia the prey concentrations are much lower and male tiger territories range in size from 800 to 1200 km^2 (320 to 480 km^2). Seasonality in terms of prey migrations, food availability and weather may also affect prey populations and therefore the size of tiger territories. Males have larger territories than females. An adult male's territory will usually overlap several females' territories. The larger area contains more than enough food, water and shelter resources, but is larger to accommodate more females' territories. Therefore, females are the most coveted resource for males. Aggression amongst adult male tigers can be influenced by the number of tigers in a given area (density) and whether there is a social disruption in which males are competing to take control of a territory. The intensity of aggression increases when there are high tiger densities for a given area because there is more competition of resources and mating opportunities. Resident male territory-holders may be challenged by other young males for possession of the territory or the young males may challenge each for ownership if the resident male has vacated or dies. The strongest male will take possession of the territory. These times of social disruption may also cause aggression between females. Tigresses' territories are smaller than that of males but focus on vital resources required for rearing young. Tigresses usually occupy territories adjacent to or take over parts of their mother's territory.

DAILY ACTIVITY CYCLE

Tigers are mainly active at night and less active during the mid-day heat. However, this pattern may vary by season and prey activity.Grooming is an important part of the tiger's day. They use their rasping tongue to remove loose hairs and dirt from their fur. The grooming process keeps the tiger's coat in good condition by using their tongues to spread oils secreted from their glands. Tigers, unlike many other cat species, readily enter water to cool themselves and in the pursuit of prey. They are powerful swimmers and capable of traversing lakes and rivers. They assert and maintain their control over their territories by continuously patrolling them. Tigers coexist with other predators such as leopards, Asiatic wild dogs, brown bears and wolves throughout most of their range. Usually there is little interaction between species especially since tigers are mostly nocturnal (active at night) and the other species are mainly diurnal (active during the day).

REPRODUCTION

Females reach sexual maturity around 3 to 4 years of age and males mature at about 4 to 5 years of age. A female tiger may enter estrus (the time when a female is receptive and capable of conceiving young) every three to nine weeks, and her receptivity lasts three to six days. In tropical climates, females may come into estrus throughout the year, though mating seems to be more frequent during the coolest months (November to April). In temperate regions, females enter estrus and mate only during the winter months.

Females advertise their readiness to mate. A few days before she enters estrus, the female will scent-mark her range more frequently with a distinctive smelling urine. During estrus, the female may frequently vocalize throughout the day to attract a male. Tigers usually begin their courtship by circling each other and vocalizing. Copulation is brief and repeated frequently for five or six days. Female tigers are induced ovulators, which means the act of mating causes the female to release an egg for fertilization. Several days of mating interactions may be required to stimulate ovulation and guarantee fertilization of the egg. Both male and female tigers may have several mates over their lifetime. The tiger's gestation period is about three and half months. It is difficult to identify a pregnant tigress because they do not begin to show a bulge until the last 10 to 12 days of pregnancy. The tigress spends the last few days of her pregnancy searching for a safe birthing place that provides enough cover to conceal the newborn cubs and has adequate prey. Each litter may have up to seven cubs, but the average is three. Tigresses usually wait between 18 to 24 months between births. Tiger cubs are born blind and are completely dependent on their mother. Newborn tiger cubs weigh between 785 and 1,610 grams (1.75 to 3.5 lb). The tiger cubs' eyes will open sometime between six to twelve days. However, they do not have their full vision for a couple of weeks.

DEVELOPMENT

Tigresses are overly cautious and secretive when caring for young cubs. She will immediately move them if the area becomes disturbed or threatened. The tigress is solely responsible for the protection and care of her young for the first few months of the cubs' lives. She leaves her young for only short periods of time to drink and hunt. Tigresses will spend nearly 70% of their time nursing their cubs for the first few days following birth. The amount of time spent nursing reduces to about 30% of their day by the time the cubs are a month old. Nursing tigresses must increase their nutritional intake by an estimated 50% to keep up their milk supply. For example a nursing female in Chitwan consumed a large prey item every five to six days as opposed to eating one large prey item every eight days when she was on her own. The tigress stimulates the cubs circulation and bowel movements by spending large periods of time licking them. The tigress may also eat the cubs' feces in order to protect them from potential predators detecting their scents. The cubs begin consuming solid food when they are six to eight weeks old. At four months of age tiger cubs are about the size of a medium-sized dog and spend their day playing, pouncing and wrestling with siblings. Tiger cubs are weaned from their mothers by six months of age. However, they are still dependent on the prey their mothers procure for them. Although they are hunting on their own yet, cubs begin to explore and roam their surroundings more freely. Male tiger cubs weigh about 90 to 105 pounds by six months of age and females are about 30 pounds lighter. Cubs will begin to follow their mothers out of the den around two months of age. However, they do not participate in the hunt at this point. They wait in a safe place for their mother to bring the food back to them. Tiger cubs begin to hunt with their mother and siblings between eight and ten months of age. The tigress is primarily concerned with teaching her young how to hunt and protect themselves. Tiger cubs spend the majority of their time playing with their siblings and their mother around fifteen months of age. Playing helps the growing tiger cubs develop useful life skills such as stalking, pouncing, swatting and climbing. A hierarchical order is established among tiger cubs by sixteen months of age with the most dominant sibling eating and consuming most resources first. The dominant cub is most often a male and will leave the family unit within a few months. Young tigers become independent from their mothers around seventeen to twenty-four months of age. Males travel further from their mother's home range than females. Young male tigers will continue to grow and develop muscle until they are about five years old but settle only temporarily in marginal habitats until they are strong enough to take a permanent territory of their own.

HUMAN IMPACT

The main threats to tiger populations today are habitat loss/fragmentation and poaching. Habitat Loss and fragmentation occurs when land is modified for agricultural purposes, logging, and land conversion for grazing domestic animals. The rapidly growing human population has reduced the number of viable tiger habitats. The human population in India alone has increased by nearly 50% since 1973 with a total population in 1995 estimated to be about 931 million. Prime tiger habitats, such as forests and grasslands, are being converted for agricultural needs.Between 1980 and 1990 in Asia, about 470,000 square km (181,467 square mi) of forest were lost. It is estimated that deforestation will continue at a rate of 47,000 square km (18,147 square mi) per year. Tigers require large interconnected tracts of suitable habitat to maintain healthy breeding populations. The conversion of land for agricultural purposes creates wide expanses of open land in which may isolate tiger populations from one another. In addition to the reduced genetic variability, fragmentation may also lead to more aggressive encounters between tigers due to the increased competition for resources and mates. Poaching is the illegal killing of an animal. Tigers are poached for two main reasons: their threat or perceived threat to wildlife and people and monetary gain. Historically tigers were poached for furs. While there is still some sold illegally, increased public awareness campaigns and international trade controls have reduced this demand. Tigers may prey upon agricultural animals and have been illegally shot at or poisoned by consuming baited carcasses. However, tigers are mainly poached for their bones and other body parts which are in great demand for traditional Chinese medicines. See Tiger Medicine section below. Illegal trade commerce is difficult to control because poaching networks are well organized and countries in which tigers live often do not have resources available to hire, equip and train law enforcement officers.

TIGER MEDICINE

Traditional Chinese medicines have utilized tiger bones for thousands of years because it is thought to calm fright and cure ulcers, bites, rheumatism, convulsions and burns. Over 110 pharmaceutical factories in 1985 were producing medicines with tiger components. The value of tiger bone varies by locality, however it is estimated that poachers receive about $130 per kilogram (2.2 pounds) in Nepal, $130 to $175 per kilogram in Vietnam and as much as $300 per kilogram in Russia. It is estimated that one complete tiger skeleton is valued at ten years worth of salary in seven nations within the tiger's range. This high demand has made tiger bones more valuable than their skin.

CHAPTER 3
CHEETAH

CLASSIFICATION
Kingdom: Animalia, Phylum: Chordata, Class: Mammalia, Order: Carnivora, Family: Felidae, Genus: *Acinonyx*, Species: *jubatus*, Sub-species: *venaticus*

GENERAL CHARACTERS
The **Asiatic cheetah** (*Acinonyx jubatus venaticus*), also known as the **Iranian cheetah**, is a critically endangered cheetah subspecies surviving today only in Iran. It used to occur in India as well, where it is locally extinct. The Asiatic cheetah lives mainly in Iran's vast central desert in fragmented pieces of remaining suitable habitat. Although once common, the cheetah was driven to extinction in other parts of Southwest Asia from Arabia to India including Afghanistan. As of 2013, only 20 cheetahs were identified in Iran but some areas remained to be surveyed. The total population is estimated to be 40 to 70 individuals, with road accidents accounting for 40% of deaths. Efforts to stop the construction of a road through the core of the Bafq Protected Area were unsuccessful. In order to raise international awareness for the conservation of the Asiatic cheetah, an illustration was used on the jerseys of the Iran national football teamate the 2014 FIFA World Cup. Currently in 2015, it is estimated that approximately 50 cheetahs are living in the wild of Iran, however their numbers are rising. The Asiatic cheetah separated from its African relative between 32,000 and 67,000 years ago. Along with the Eurasian lynx and the Persian leopard, it is one of three remaining species of large cats in Iran today. During the British colonial times in India it was called **hunting leopard**, a name derived from the ones that were kept in captivity in large numbers by the Indian royalty to use in hunting wild antelopes. In Dutch, the cheetah is still called *jachtluipaard*. The Hindi word cītā is derived from the Sanskrit word *chitraka* meaning "speckled". Asiatic cheetahs are slimmer, lighter and slightly shorter than their African brethren. The head and

body of an adult Asiatic cheetah measure from 112–135 cm (44–53 in) with a tail length between 66 and 84 cm (26 and 33 in). It weighs from 34 to 54 kg (75 to 119 lb). Males are slightly larger than the females.The cheetah is the fastest land animal in the world. It was previously thought that the body temperature of a cheetah increases during a hunt due to high metabolic activity. In a short period of time during a chase, a cheetah may produce 60 times more heat than at rest, with much of the heat, produced from glycolysis, stored to possibly raise the body temperature. The claim was supported by data from experiments in which two cheetahs ran on a treadmill for minutes on end but contradicted by studies in natural settings, which indicate that body temperature stays relatively the same during a hunt. A 2013 study suggested stress hyperthermia and a slight increase in body temperature after a hunt. The cheetah's nervousness after a hunt may induce stress hyperthermia, which involves high sympathetic nervous activity and raises the body temperature. After a hunt, the risk of another predator taking their kill is great and the cheetah is on high alert and stressed. The increased sympathetic activity prepares the cheetah's body to run when another predator approaches. In the 2013 study, even the cheetah that did not chase the prey experienced an increase in body temperature once the prey was caught, showing increased sympathetic activity.

HABIT AND HABITAT
The Asiatic cheetah preys on small antelopes. In Iran, its diet consists mainly of Jebeer gazelle (also called Chinkara), Goitered gazelle, wild sheep, wild goat, and Cape hare. The main threat to the species is loss of their primary prey species due to poaching and grazing competition with domestic livestock. A study published in 2012 indicated that hares and rodents, while forming part of the cheetah's diet, are not a significant source of nutrition due to their small size and difficulty of being caught. In India, prey was formerly abundant. Before its extinction in the country, the cheetah fed on the blackbuck, the chinkara, and sometimes the chital and the nilgais in low, isolated, rocky hills, near the plains on which live antelopes, its principal prey. It also kills gazelles, nilgai, and, doubtless, occasionally deer and other animals. Instances also occur of sheep and goats being carried off by it, but it rarely molests domestic animals, and has not been known to attack men. Its mode of capturing its prey is to stalk up to within a moderate distance of between one to two hundred yards, taking advantage of inequalities of the ground, bushes, or other cover, and then to make a rush. Its speed for a short distance is remarkable far exceeding that of any other beast of prey, even of a greyhound or kangaroo-hound, for no dog can at first overtake an Indian antelope or a gazelle, either of which is quickly run down by *C. jubatus*, if the start does not exceed about two hundred yards. General Mc Master saw a very fine hunting-leopard catch a black buck that had about that start within four hundred yards. It is probable that for a short distance the hunting-leopard is the swiftest of all mammals.Cheetahs thrive in open lands, small plains, semi-desert areas, and

other open habitats where prey is available. The Asiatic cheetah is found mainly in the desert areas around Dasht-e Kavir in the eastern half of Iran, including parts of the Kerman, Khorasan, Semnan,Yazd, Tehran, and Markazi provinces. Most live in five sanctuaries: Kavir National Park, Touran National Park, Bafq Protected Area, Daranjir Wildlife Reserve, and Naybandan Wildlife Reserve. Remaining cheetahs are divided into widely separated populations. Some possibly survive in the dry open Balochistan province of Pakistan but locals said they had not seen it for more than fifteen years. During the 1970s, cheetahs in Iran were estimated to number about 200 individuals in seven protected areas. Continuous field surveys, along with 12,000 nights of camera trapping, were used to estimate the population size. Using 80 camera traps placed throughout the Dasht-e Kavir plateau, Iranian researchers obtained images of 76 individual cheetahs over the course of ten years from 2001. In December 2014, four cheetahs were sighted and photographed by camera traps in the Touran National Park. Asiatic cheetahs once ranged from the Arabian Peninsula to India, through Iran, Central Asia, Afghanistan, and Pakistan. In India, Asiatic cheetahs occurred in Rajputana, Punjab, Sind, and south of the Ganges from Bengal to the northern part of theDeccan Plateau. Asiatic cheetahs were also found in other parts of India including the Kaimur District(present-day eastern Uttar Pradesh, near Bihar), Darrah and other desert regions of Rajasthan and parts of Gujarat and Central India. A female was sighted in the Koriya district in 1951. In Afghanistan, Asiatic cheetahs are considered extinct since the 1950s. Uncontrolled hunting of Asiatic cheetahs and their prey, severe winters and conversion of grassland to agriculturally used areas contributed to the decline of cheetahs in Central Asia. The last reported sighting in Uzbekistan dates end of 1983. The last reported killed in Turkmenistandates November 1984.

REPRODUCTION

Evidence of mothers successfully raising cubs is very rare. In May 2013, images from a camera trap showed a mother with three cubs aged approximately one year in Miandasht Wildlife Refuge in north-east Iran. In October 2013, conservationists from the Persian Wildlife Heritage Foundation filmed a mother with four cubs in Touran. On 7 January 2015, Director General of Environmental Protection Department in North Khorasan, Iran announced a sighting of a female Asiatic cheetah and her cub at Miandasht Wildlife Refuge. Motahari also maintained that two days prior to this sighting, three other adult cheetahs were sighted by the locals some kilometers to the eastern border of Miandasht, and immediately reported to Jajrom Department of Environment.

THREATS

Reduced gazelle numbers, persecution, land-use change, habitat degradation and fragmentation, and desertification contributed to the cheetah's decline. According to the Iranian Department of Environment this degradation occurred mainly between 1988 and 1991. The cheetah is affected by loss of prey as a result of overgrazing from introduced livestock and antelope hunting. Its prey was pushed out as herders entered game reserves with their herds. Mining development and road construction near reserves also threaten the population. Coal, copper, and iron have been mined in the cheetah's habitat in three different regions in central and eastern Iran. It is estimated that the two regions for coal (Nayband) and iron (Bafq) have the largest cheetah population outside the protected areas. Mining itself is not a direct threat to cheetahs; road construction and the resulting traffic have made the cheetah accessible to humans, including poachers. The Iranian border regions to Afghanistan and Pakistan (Baluchistan province) are major passages for armed outlaws and opium smugglers who are active in the central and western regions of Iran, passing through cheetah habitat. According to Asada in 1997, the region suffers from uncontrolled hunting throughout the desert and the governments of the three countries cannot establish control. There is no reliable information regarding the present situation in this region. In February 2015, it was reported that road accidents were responsible for 40% of deaths.

CONSERVATION

The Asiatic cheetah is now listed as **critically endangered** in the IUCN Red List of Threatened Animals. Following the Iranian Revolution of 1979, wildlife conservation was given a lower priority, but in recent years Iran has made efforts to conserve the remaining population. Iran's Department of the Environment, the United Nations Development Programme (UNDP), and the Global Environment Facility (GEF) have launched the Conservation of the Asiatic Cheetah Project (CACP) designed to preserve and rehabilitate the remaining areas of cheetah habitat left in Iran. Some surveys by Asadi in the latter half of 1997 showed that urgent action was required to rehabilitate wildlife populations, especially gazelles and their habitat, if the Asiatic cheetah is to survive. The Wildlife Conservation Society (WCS) and the Department of Environment, Iran (DoE) began a collaring project for Asiatic cheetahs in the fall of 2006. GPS collars provide data on the cat's movements. International sanctions have made some projects, such as obtaining camera traps, difficult. In 2006, Iran designated 31 August as the Cheetah Conservation Day, during which the public is informed about conservation programs. In 2013, it was reported that the cheetah might appear on the Iranian national football team's jerseys at the 2014 FIFA World Cup. FIFA approved the design on 1 February 2014.

CHAPTER 4
LION

Female (lioness)   Male

CLASSIFICATION
Kingdom: Animalia, Phylum: Chordata, Class: Mammalia, Order: Carnivora, Family: Felidae, Genus: *Panthera***, Species:** *leo*

GENERAL CHARACTERS
Distribution of lions in India: The Gir Forest, in Gujarat, is the last natural range of about 400 wild Asiatic lions. There are plans to reintroduce some lions to Kuno Wildlife Sanctuary in neighbouring Madhya Pradesh. The **lion** (*Panthera leo*) is one of the five big cats in the genus *Panthera* and a member of the family Felidae. The commonly used term **African lion** collectively denotes the several subspecies found in Africa. With some males exceeding 250 kg (550 lb) in weight, it is the second-largest living cat after thetiger. Wild lions currently exist in sub-Saharan Africa and inAsia (where an endangered remnant population resides in Gir Forest National Park in India) while other types of lions have disappeared from North Africa and Southwest Asia in historic times. Until the late Pleistocene, about 10,000 years ago, the lion was the most widespread large land mammal afterhumans. They were found in most of Africa, across Eurasia from western Europe to India, and in the Americas from the Yukon to Peru. The lion is a vulnerable species, having seen a major population decline in its African range of 30–50% per two decades during the second half of the 20th century. Lion populations are untenable outside designated reserves and national parks. Although the cause of the decline is not fully understood, habitat loss and conflicts with humans are currently the greatest causes of concern. Within Africa, the West African lion population is particularly endangered. Behind only the tiger, the lion is the second largest living felid in length and weight. Its skull is very similar to that of the tiger, although the frontal region is usually more

depressed and flattened, with a slightly shorter postorbital region. The lion's skull has broader nasal openings than the tiger, however, due to the amount of skull variation in the two species, usually, only the structure of the lower jaw can be used as a reliable indicator of species. Lion colouration varies from light buff to yellowish, reddish, or dark ochraceous brown. The underparts are generally lighter and the tail tuft is black. Lion cubs are born with brown rosettes (spots) on their body, rather like those of a leopard. Although these fade as lions reach adulthood, faint spots often may still be seen on the legs and under parts, particularly on lionesses. Lions are the only members of the cat family to display obvious sexual dimorphism – that is, males and females look distinctly different. They also have specialized roles that each gender plays in the pride. For instance, the lioness, the hunter, lacks the male's thick mane. The colour of the male's mane varies from blond to black, generally becoming darker as the lion grows older. The most distinctive characteristic shared by both females and males is that the tail ends in a hairy tuft. In some lions, the tuft conceals a hard "spine" or "spur", approximately 5 mm long, formed of the final sections of tail bone fused together. The lion is the only felid to have a tufted tail – the function of the tuft and spine are unknown. Absent at birth, the tuft develops around 5½ months of age and is readily identifiable at 7 months. The size of adult lions varies across their range with those from the southern African populations in Zimbabwe, the Kalahari and Kruger Park averaging around 189.6 kg (418 lb) and 126.9 kg (280 lb) in males and females respectively compared to 174.9 kg (386 lb) and 119.5 kg (263 lb) of male and female lions from East Africa. Reported body measurements in males are head-body lengths ranging from 170 to 250 cm (5 ft 7 in to 8 ft 2 in), tail lengths of 90–105 cm (2 ft 11 in–3 ft 5 in). In females reported head-body lengths range from 140 to 175 cm (4 ft 7 in to 5 ft 9 in), tail lengths of 70–100 cm (2 ft 4 in–3 ft 3 in), however, the frequently cited maximum head and body length of 250 cm (8 ft 2 in) fits rather to extinct Pleistocene forms, like the American lion, with even large modern lions measuring several centimetres less in length. Record measurements from hunting records are supposedly a total length of nearly 3.6 m (12 ft) for a male shot near Mucsso, southern Angola in October 1973 and a weight of 313 kg (690 lb) for a male shot outside Hectorspruit in eastern Transvaal, South Africa in 1936. Another notably outsized male lion, which was shot near Mount Kenya, weighed in at 272 kg (600 lb).

HABIT AND HABITAT
Lions prefer to scavenge when the opportunity presents itself with carrion providing more than 50% of their diet. They scavenge animals either dead from natural causes (disease) or killed by other predators, and keep a constant lookout for circling vultures, being keenly aware that they indicate an animal dead or in distress. In fact, most dead prey on which both hyenas and lions feed upon are killed by the hyenas instead of the lions.

The prey consists mainly of medium-sized mammals, with a preference for wild beast, zebras, buffalo, and warthogs in Africa and nilgai, wild boar, and several deer species in India. Many other species are hunted, based on availability, mainly ungulates weighing between 50 and 300 kg (110 and 660 lb) such as kudu, hartebeest, gemsbok, and eland. Occasionally, they take relatively small species such as Thomson's gazelle or springbok. Lions hunting in groups are capable of taking down most animals, even healthy adults, but in most parts of their range they rarely attack very large prey such as fully grown male giraffes due to the danger of injury. Giraffes and buffaloes are almost invulnerable to a solitary lion as well.

Extensive studies show that lionesses normally prey on mammals with an average weight of 126 kg (278 lb), while kills made by male lions average 399 kg (880 lb). In Africa, wildebeest rank at the top of preferred prey (making nearly half of the lion prey in the Serengeti) followed by zebra. Lions do not prey on fully grown adult elephants; most adult hippopotamuses, rhinoceroses, and smaller gazelles, impala, and other agile antelopes are generally excluded. However, giraffes and buffaloes are often taken in certain regions. For instance, in Kruger National Park, giraffes are regularly hunted. In Manyara Park, Cape buffaloes constitute as much as 62% of the lion's diet, due to the high number density of buffaloes. Occasionally hippopotamus is also taken, but adult rhinoceroses are generally avoided. Warthogs are often taken depending on availability. The lions of Savuti, Botswana, have adapted to hunting young elephants during the dry season, and a pride of 30 lions has been recorded killing individuals between the ages of four and eleven years. In the Kalahari desert in South Africa, black-maned lions may chase baboons up a tree, wait patiently, then attack them when they try to escape:

Lions also attack domestic livestock and in India cattle contribute significantly to their diet. Lions are capable of killing other predators such as leopards, cheetahs, hyenas, and wild dogs, though (unlike most felids) they seldom devour the competitors after killing them. A lion may gorge itself and eat up to 30 kg (66 lb) in one sitting; if it is unable to consume all the kill it will rest for a few hours before consuming more. On a hot day, the pride may retreat to shade leaving a male or two to stand guards An adult lioness requires an average of about 5 kg (11 lb) of meat per day, a male about 7 kg (15 lb). In Africa, lions can be found in savanna grasslands with scattered *Acacia* trees, which serve as shade; their habitat in India is a mixture of dry savanna forest and very dry deciduous scrub forest. The habitat of lions originally spanned the southern parts of Eurasia, ranging from Greece to India, and most of Africa except the central rainforest-zone and the Sahara desert. The species was eradicated from Palestine by the middle Ages and from most of the rest of Asia after the arrival of readily available firearms in the eighteenth century. Between the late nineteenth and early twentieth century, they became extinct in North Africa and Southwest Asia. By the late nineteenth century, the lion had disappeared from Turkey and most of northern India,while the last sighting of a live Asiatic lion in Iran was in 1941 (between Shiraz and Jahrom, Fars Province), although the corpse of a lioness was found on the banks of the Karun river, Khūzestān Province in 1944. There are no subsequent reliable reports from Iran.The subspecies now survives only in and around the Gir Forest of northwestern India. Approximately 400 lions live in the area of the 1,412 km^2 (545 sq mi) sanctuary in the state of Gujarat, which covers most of the forest. Their numbers have increased from 180 to 411 animals mainly because the natural prey species have recovered.

SPECIAL CHARACTERS

The mane of the adult male lion, unique among cats, is one of the most distinctive characteristics of the species. It may provide an excellent intimidation display; aiding the lion during confrontations with other lions. The presence, absence, colour, and size of the mane is associated with genetic precondition, sexual maturity, climate, and testosterone production; the rule of thumb is the darker and fuller the mane, the healthier the lion. Sexual selection of mates by lionesses favors males with the densest, darkest mane. Research inTanzania also suggests mane length signals fighting success in male–male relationships. Darker-maned individuals may have longer reproductive lives and higher offspring survival, although they suffer in the hottest months of the year. In the Pendjari National Park area almost all males are maneless or have very weak manes. Maneless male lions have also been reported from Senegal, from Sudan (Dinder National Park), and from Tsavo East National Park in Kenya, and the original male white lion from Timbavati also was maneless. The testosterone hormone has been linked to mane growth, therefore castrated lions often have minimal to no mane, as the removal of the gonads inhibits testosterone production. Cave paintings of extinct European cave lions almost exclusively show animals with no manes, suggesting that either they were maneless, or that the paintings depict lionesses as seen hunting in a group.

White lion

White lions owe their colouring to a recessive allele; they are rare forms of the subspecies *Panthera leo krugeri*.

The white lion is not a distinct subspecies, but a special morph with a genetic condition, leucism that causes paler colouration akin to that of the white tiger; the condition is similar to melanism, which causes black panthers. They are not albinos, having normal pigmentation in the eyes and skin. White Transvaal lion (*Panthera leo krugeri*) individuals occasionally have been encountered in and around Kruger National Park and the adjacent Timbavati Private Game Reserve in eastern South Africa, but are more commonly found in captivity, where breeders deliberately select them. The unusual cream colour of their coats is due to a recessive allele. Reportedly, they have been bred in camps in South Africa for use as trophies to be killed during canned hunts.

REPRODUCTION

Most lionesses will have reproduced by the time they are four years of age. Lions do not mate at any specific time of year, and the females are polyestrous. As with other cats' penises, the male lion's penis has spines that point backward. During withdrawal of the penis, the spines rake the walls of the female's vagina, which may cause ovulation. A lioness may mate with more than one male when she is in heat.The average gestation period is around 110 days, the female giving birth to a litter of one to four cubs in a secluded den (which may be a thicket, a reed-bed, a cave, or some other sheltered area) usually away from the rest of the pride. She will often hunt by herself while the cubs are still helpless, staying relatively close to the thicket or den where the cubs are kept. The cubs themselves are born blind – their eyes do not open until roughly a week after birth. They weigh 1.2–2.1 kg (2.6–4.6 lb) at birth and are almost helpless, beginning to crawl a day or two after birth and walking around three weeks of age. The lioness moves her cubs to a new den site several times a month, carrying them one by one by the nape of the neck, to prevent scent from building up at a single den site and thus avoiding the attention of predators that may harm the cubs.Usually, the mother does not integrate herself and her cubs back into the pride until the cubs are six to eight weeks old. Sometimes this introduction to pride life occurs earlier, however, particularly if other lionesses have given birth at about the same time. For instance, lionesses in a pride often synchronise their reproductive cycles so that they cooperate in the raising and suckling of the young (once the cubs are past the initial stage of isolation with their mother), who suckle indiscriminately from any or all of the nursing females in the pride. In addition to greater protection, the synchronization of births also has an advantage in that the cubs end up being roughly the same size, and thus have an equal chance of survival. If one lioness gives birth to a litter of cubs a couple of months after another lioness, for instance, then the younger cubs, being much smaller than their older brethren, usually are dominated by larger cubs at mealtimes – consequently, death by starvation is more common among the younger cubs. In addition to starvation, cubs also face many other dangers, such as predation by jackals, hyenas, leopards, martial eagles, and snakes. Even buffaloes, should they catch the scent of lion cubs, often stampede toward the thicket or den where they are being kept, doing their best to trample the cubs to death while warding off the lioness. Furthermore, when one or more new males oust the previous male(s) associated with a pride, the conqueror(s) often kill any existing young cubs, perhaps because females do not become fertile and receptive until their cubs mature or die. Among all, about 80% of the cubs will die before the age of two. When first introduced to the rest of the pride, the cubs initially lack confidence when confronted with adult lions other than their mother. They soon begin to immerse themselves in the pride life, however, playing among themselves or attempting to initiate play with the adults. Lionesses with cubs of their own are more likely to be tolerant of another lioness's cubs than lionesses without cubs. The tolerance of the male lions toward the cubs varies – sometimes, a male will patiently let the cubs play with his tail or his mane, whereas another may snarl and bat the cubs away. Male lions are generally more likely to share food caught by the pride with cubs than with lionesses, but rarely share their own catches with others. Weaning occurs after six to seven months. Male lions reach maturity at about 3 years of age and, at 4–5 years of age, are capable of challenging and displacing the adult male(s) associated with another pride. They begin to age and weaken between 10 and 15 years of age at the latest, if they have not already been critically injured while defending the pride (once ousted from a pride by rival males, male lions rarely manage a second take-over). This leaves a short window for their own offspring to be born and mature. If they are able to procreate as soon as they take over a pride, potentially, they may have more offspring reaching maturity before they also are displaced. A lioness often will attempt to defend her cubs fiercely from a usurping male, but such actions are rarely successful. He usually kills all of the existing cubs who are less than two years old. A lioness is weaker and much lighter than a male; success is more likely when a group of three or four mothers within a pride join forces against one male.

CHAPTER 5
NILGAI

CLASSIFICATION
Kingdom: Animalia, Phylum: Chordata, Class: Mammalia, Order: Artiodactyla, Family: Bovidae, Subfamily: Bovinae, Genus: *Boselaphus,* Species: *tragocamelus*

GENERAL CHARACTERS
The **nilgai** (*Boselaphus tragocamelus*), sometimes called *nilgau*, is the largest Asian antelope. It is one of the most commonly seen wild animals of central and northern India, often seen in farmland or scrub forest. The mature male appears ox-like and is also known as the **blue bull**. A blue bull is called a *nil gai* or *nilgai* in India, from *nil* meaning blue and *gai* meaning a bovine animal (literally 'cow'). It is also present in parts of southern Nepal and eastern Pakistan. The species has become extinct in Bangladesh. It was known as the *nilghor* (*nil* = blue, *ghor* = horse) during the rule of Aurangzeb in the Mughal era. It is the only member of genus ***Boselaphus***. Nilgai stand 1.1 to 1.5 m (3 ft 7 in to 4 ft 11 in) at the shoulder and measure 1.7 to 2.1 m (5 ft 7 in to 6 ft 11 in) in head-body length, with a 45- to 50-cm (18- to 20-in) tail. Males are larger than females, weighing 109 to 288 kg (240 to 635 lb), with a maximum of 308 kg (679 lb), compared with the adult female weight of around 100 to 213 kg (220 to 470 lb). The nilgai has thin legs and a robust body that slopes down from the shoulder. They show marked sexual dimorphism, with only the males having horns. Adult males have a grey to bluish-grey coat, with white spots on the cheeks and white colouring on the edges of the lips. They also have a white throat bib and a narrow white stripe along the underside of the body that widens at the rear. The tips of the long, tufted tail and of the ears are black. They also possess a tubular-shaped "pennant" of long, coarse hair on the midsection of the throat. The males have two black, conical horns, arising close together just behind the eyes. The horns project upwards, but are slightly curved forward; they measure between 15 and 24 centimeters (5.9 and 9.4 in) in a fully grown adult. Although the horns are usually smooth, in some older males, they may

develop ring-shaped ridges near the base. In contrast, females and young are tawny brown in colour, although otherwise with similar markings to the male; they have no horns and only a very small "pennant". Both sexes have an erect mane on the back of the neck, terminating in a bristly "hog-tuft" just above the shoulders.

HABIT AND HABITAT

Nilgai antelope are found throughout most of India, from the base of the Himalayas in the north, down to the state of Karnataka in the south, being absent only in eastern Bengal, Assam, the Malabar Coast, and regions close to the Bay of Bengal. They inhabit the Gir forest and across Rajasthan in the west to the states of Assam and West Bengal in the east. In Nepal, they occur patchily in the southern lowlands. Nilgai are habitat generalists, living in grasslands and woodlands where they eat grasses, leaves, buds, and fruit.

REPRODUCTION

Breeding occurs in late autumn to early winter. Prior to therut, males compete to establish dominance. Males display to each other by holding their heads erect and presenting the white patch and tassel on their throats. They may also rush towards one another, holding their heads down so their horns project forwards. Such displays often escalate to direct conflict, including head-butting and neck-fighting. Although bulls have thick skin on their heads and necks, which helps protect them in such fights, serious injury can nonetheless occur. Females also compete to establish dominance around the time of the rut, including neck-fighting, and butting rivals on their shoulders or flanks. Males mate with several females over the course of the breeding season, but do not establish clear harems, instead wandering between different all-female groups. Courtship lasts about 45 minutes, with the male adopting a stiff gait with tail held erect, and the female responding with a flehmen gesture and raised tail before permitting mounting. Gestation lasts 243 to 247 days, resulting in the birth of twins in about 50% of cases, although births of one or three do occur. Females become solitary towards the end of their pregnancy, and hide their young from other nilgai for the first month of their lives. The calves are precocious, being able to stand within 40 minutes of birth, and they begin to forage during their fourth week of life. Calves usually weigh 14 to 16 kg (31 to 35 lb) at birth.Females reach sexual maturity at around two years of age, and males by their third year, although the most reproductively active bulls are typically at least four or five years old. They live for around 12 or 13 years in the wild, but have survived for up to 21 years in captivity.

CHAPTER 6
BISON

CLASSIFICATION
Kingdom: Animalia, Phylum: Chordata, Class: Mammalia, Order: Artiodactyla, Family: Bovidae, Subfamily: Bovinae, Genus: *Bos,* Species: *gaurus*

GENERAL CHARACTERS
The **Indian bison,** also called **gaur (*Bos gaurus*)**, is the largest extant bovine and is native to South Asia and Southeast Asia. The species is listed as **vulnerable** on the IUCN Red List since 1986, as the population decline in parts of the species' range is likely to be well over 70% during the last three generations. Population trends are stable in well-protected areas, and are rebuilding in a few areas which had been neglected. The gaur is the tallest species of wild cattle. The Malayan gaur is called *seladang*, and the Burmese gaur is called *pyoung*. The domesticated form of the gaur *Bos frontalis* is called *gayal* or *mithun*. The Gaur is a strong and massively built species with a high convex ridge on the forehead between the horns, which bends forward, causing a deep hollow in the profile of the upper part of the head. There is a prominent ridge on the back. The ears are very large; the tail only just reaches the hocks, and in old bulls the hair becomes very thin on the back. In colour, the adult male gaur is dark brown, approaching black in very old individuals; the upper part of the head, from above the eyes to the nape of the neck, is, however, ashy grey, or occasionally dirty white; the muzzle is pale coloured, and the lower part of the legs are pure white or tan. The cows and young bulls are paler, and in some instances have a rufous tinge, which is most marked in groups inhabiting dry and open districts. The tail is shorter than in the typical oxen, reaching only to the hocks. They have a distinct ridge running from the shoulders to the middle of the back; the shoulders may be as much as 12 cm (4.7 in) higher than the rump. This ridge is caused by the great length of the spinous processes of the vertebrae of the fore-part of the trunk as compared with those of the lions. The hair is short, fine and glossy, and the hooves are narrow and pointed. The gaur has a head-and-body length of 250 to 330 cm (8 ft 2 in to 10 ft 10 in) with a 70 to 105 cm (28 to 41 in)

long tail, and is 165 to 220 cm (5 ft 5 in to 7 ft 3 in) high at the shoulder. The average weight of adult gaur is 650 to 1,000 kg (1,430 to 2,200 lb), with an occasional large bull weighing up to 1,500 kg (3,300 lb). Males are about one-fourth larger and heavier than females. In general measurements are derived from gaurs surveyed in India and China. The Seladang, or Malayasian subspecies, may average larger but no scientifically published measurements are known. Gaur do not have a distinct dewlap on the throat and chest. Both sexes carry horns, which grow from the sides of the head, curving upwards. Between the horns is a high convex ridge on the forehead. At their bases they present an elliptical cross-section, a characteristic that is more strongly marked in bulls than in cows. The horns are decidedly flattened at the base and regularly curved throughout their length, and are bent inward and slightly backward at their tips. The colour of the horns is some shade of pale green or yellow throughout the greater part of their length, but the tips are black. The horns, of medium size by large bovid standards, grow to a length of 60 to 115 cm (24 to 45 in). The cow is considerably lighter in make and in colour than the bull. The horns are more slender and upright, with more inward curvature, and the frontal ridge isscarcely perceptible. In young animals the horns are smooth and polished. In old bulls they are rugged and dented at the base. Gaurs are among the largest living land animals. Only elephants, rhinos, the hippopotamus (*Hippopotamus amphibius*) and the giraffe (*Giraffa camelopardalis*) consistently grow heavier. Two species that naturally co-exist with the gaur are heavier: the Asian elephant(*Elephas maximus*) and Indian rhinoceros (*Rhinoceros unicornis*). By most standards of measurements, gaur is the largest wild bovid alive today. However, the shorter-legged, bulkier Wild water buffalo (*Bubalus arnee*) is similar in average body mass, if not maximum weight.

HABIT AND HABITAT

Bison are **herbivores** and eat simple foods. The bison's main foodstuff is grass, though they will eat any available low-lying shrubbery, as well as sedges. In the winter, bison forage for grass under the snow. If little grass is available, they will eat the twigs of shrubs. Bison are notably better browsers than cattle, since cattle are more obligate grazers, though wood bison have also been described as "obligate grazers".Wisent tend to browse on shrubs and low-hanging trees more often than do the American bison, which prefer grass to shrubbery and trees.Gaur historically occurred throughout mainland South and Southeast Asia, including Vietnam, Cambodia, Laos,China, Thailand, Peninsular Malaysia, Myanmar, India,Bangladesh, Bhutan, and Nepal. Today, the species is seriously fragmented within its range, and regionally extinct in Sri Lanka. Gaur are largely confined to evergreen forests or semi-evergreen and moist deciduous forests, but also occur in deciduous forest areas at the periphery of their range. Gaur habitat is characterized by large, relatively undisturbed forest tracts, hilly terrain below an altitude of 5,000 to 6,000 ft (1,500 to

1,800 m),availability of water, and an abundance of forage in the form of grasses,bamboo, shrubs, and trees. Their apparent preference for hilly terrain may be partly due to the earlier conversion of most of the plains and other low-lying areas to croplands and pastures. They occur from sea level to an altitude of at least 2,800 m (9,200 ft). Low-lying areas seem to comprise optimal habitat. In India, the population was estimated to be 12,000–22,000 in the mid-1990s. The Western Ghats and their outflanking hills in southern India constitute one of the most extensive extant strongholds of gaur, in particular in the Wayanad – Nagarhole – Mudumalai –Bandipur complex. The populations in India, Bhutan and Bangladesh are estimated to comprise 23,000–34,000 individuals. Major populations of about 2,000 individuals have been reported in both Nagarahole and Bandipur National Parks, over 1,000 individuals in Tadoba Andhari Tiger Project, 500–1000 individuals in both Periyar Tiger Reserve and Silent Valley and adjoining forest complexes, and over 800 individuals in Bhadra Wildlife Sanctuary.

SPECIAL CHARACTERS

Wallowing is a common behavior of bison. A bison wallow is a shallow depression in the soil, either wet or dry. Bison roll in these depressions, covering themselves with mud or dust. Possible explanations suggested for wallowing behavior include grooming behavior associated with moulting, male-male interaction (typically rutting behavior), social behavior for group cohesion, play behavior, relief from skin irritation due to biting insects, reduction of ectoparasite load (ticks and lice), and thermoregulation. In the process of wallowing, bison may become infected by the fatal disease anthrax, which may occur naturally in the soil. Bison temperament is often unpredictable. They usually appear peaceful, unconcerned, even lazy, yet they may attack anything, often without warning or apparent reason. They can move at speeds up to 35 mph (56 km/h) and cover long distances at a lumbering gallop. Their most obvious weapons are the horns borne by both males and females, but their massive heads can be used as battering rams, effectively using the momentum produced by 2,000 pounds (900 kg) moving at 30 mph (50 km/h). The hind legs can also be used to kill or maim with devastating effect. At the time bison ran wild, they were rated second only to the Alaska brown bear as a potential killer, more dangerous than the grizzly bear. In the words of early naturalists, they were dangerous, savage animals that feared no other animal and in prime condition could best any foe(except for wolves and brown bears).The rutting, or mating, season lasts from June through September, with peak activity in July and August. At this time, the older bulls rejoin the herd, and fights often take place between bulls. The herd exhibits much restlessness during breeding season. The animals are belligerent, unpredictable, and most dangerous. Gaur historically occurred throughout mainland South and Southeast Asia, including Vietnam, Cambodia, Laos,China, Thailand, Peninsular Malaysia, Myanmar, India,Bangladesh,

Bhutan, and Nepal. Today, the species is seriously fragmented within its range, and regionally extinct in Sri Lanka. Gaur are largely confined to evergreen forests or semi-evergreen and moist deciduous forests, but also occur in deciduous forest areas at the periphery of their range. Gaur habitat is characterized by large, relatively undisturbed forest tracts, hilly terrain below an altitude of 5,000 to 6,000 ft (1,500 to 1,800 m),availability of water, and an abundance of forage in the form of grasses,bamboo, shrubs, and trees. Their apparent preference for hilly terrain may be partly due to the earlier conversion of most of the plains and other low-lying areas to croplands and pastures. They occur from sea level to an altitude of at least 2,800 m (9,200 ft). Low-lying areas seem to comprise optimal habitat. In India, the population was estimated to be 12,000–22,000 in the mid-1990s. The Western Ghats and their outflanking hills in southern India constitute one of the most extensive extant strongholds of gaur, in particular in the Wayanad – Nagarhole – Mudumalai – Bandipur complex.

REPRODUCTION

Gaur have one calf (or occasionally two) after a gestation period of about 275 days, about nine months, a few days less than domestic cattle. Calves are typically weaned after seven to 12 months. Sexual maturity occurs in the gaur's second or third year. Breeding takes place year-round, but typically peaks between December and June. The lifespan of a gaur in captivity is up to 30 years.

CHAPTER 7
ANTELOPE

CLASSIFICATION
Kingdom: Animalia, Phylum: Chordata, Class: Mammalia, Order: Artiodactyla, Family: Bovidae, Subfamily: Bovinae, Genus: *Tetracerus*, Species: *quadricornis*

GENERAL CHARACTERS
The **four-horned antelope** (*Tetracerus quadricornis*), or **chousingha**, is a species of small antelope found in open forest in India and Nepal. It is the only species currently classified in the genus *Tetracerus*. Standing only 55 to 64 cm (22 to 25 in) at the shoulder, it is the smallest of Asian bovids. Males of the species are unique among extant mammals in that they possess four permanent horns. The species is listed as **Vulnerable** by the IUCN due to habitat loss. The four-horned antelope is among the smallest Asian bovids, standing just 55 to 64 cm (22 to 25 in) tall at the shoulder, and weighing 17 to 22 kg (37 to 49 lb). It has a generally slender build, with thin legs and a short tail. The coat is yellow-brown or reddish, fading to a whitish colour on the under parts and the insides of the legs. A black stripe of hair runs down the anterior surface of each leg, with black patches on the muzzle and the backs of the ears. Females have four teats, located far back on the abdomen. The most distinctive feature of the animal is the presence of four horns; a feature unique among extant mammals. Only the males grow horns, usually with two between the ears and a second pair further forward on the forehead. The first pair of horns appears at just a few months of age, and the second pair generally grows after 10 to 14 months. The horns are never shed, although they may be damaged during fights. Not all adult males have horns; in some individuals, especially those belonging to the subspecies *T. q. subquadricornis*, the forward pair of horns is absent or represented only by small, hairless bumps. The hind pair of horns reaches 7 to 10 cm (2.8 to 3.9 in) in length, while the forward pair is usually smaller, at just 2 to 5 cm (0.79 to 1.97 in).

HABIT AND HABITAT

They are **herbivorous,** feeding on soft leaves, fruits, and flowers. Although the precise details of their diets in the wild are unknown, they have been observed to prefer plants such as Indian plum, Indian gooseberry, *Bauhinia*, and *Acacia* in artificial trials.Most wild four-horned antelope are found in India, with small, isolated populations in Nepal. Their range extends south of the Gangetic plains down to the state of Tamil Nadu, and east as far as Odisha. They also occur in the Gir Forest National Park of western India.Four-horned antelope live in a variety of habitats across their range, but prefer open, dry, deciduous forests in hilly terrain. They tend to remain in areas with significant vegetation cover from tall grasses or heavy undergrowth, and close to a supply of water. They generally stay away from human-inhabited areas.Predators of four-horned antelopes include tigers, leopards, and dholes.

Three subspecies of four-horned antelope are recognized: *T. q. quadricornis, T. q. iodes, T. q.subquadricornis*.

SPECIAL CHARACTERS

Four-horned antelope are generally solitary animals, although they are occasionally found in groups of up to four individuals. They are sedentary, rather than nomadic, and may defend exclusive territories. Males tend to become very aggressive towards other males during mating seasons. Adults make alarm calls that sound like a husky 'phronk', and other, quieter calls to communicate with young or other adults. They also communicate through scent marking, leaving piles of droppings in their territories, and marking vegetation using large scent glands in front of the eyes.

REPRODUCTION

The breeding season lasts from May to July, and males and females generally remain apart for the remainder of the year. Courtship behaviour consists of the male and female kneeling and pushing at each other with intertwined necks, followed by ritual strutting by the male. Gestationlasts about eight months, and results in the birth of one or two young. At birth, the young are 42 to 46 cm (17 to 18 in) long, and weigh 0.74 to 1.1 kg (1.6 to 2.4 lb). Young remain with the mother for about a year, and reach sexual maturity at around two years. With ninety-one species in Africa and Eurasia, antelope show quite a bit of variation in anatomy, morphology and mating behavior. The different species seem to show three distinct mating strategies. Recent studies demonstrate some very unusual trickery and fighting behaviour associated with mating among these animals.

A Little About Antelope

Antelope are members of the bovid family, along with other hooved, ruminating, horned mammals. Ranging in size from the tiny dik-dik, at 16 pounds, to the royal antelope, at 2,100, all are fast runners with slender, powerful limbs. In some species both males and females possess horns; however, these are significantly more developed in males of species among which males fight each other for females during mating season.

Monogamous Antelope

Some antelope, such as the dik-dik, are generally monogamous. This seems to less from choice and more as a result of environmental and predatory constraints. Resources in certain forested habitats are sparse, and therefore the benefits of producing and providing for more than one offspring are small. In addition, these sparse forests allow for little cover from predators, thus making wandering from a home group dangerous. These antelope may pair off, or, when resources allow, one male may have a harem of two to four females.

Lek Breeding System

The lek breeding system is more like an auction. Males will enter a territory and fight over it while females observe and choose the most competitive male. This male then has breeding rights with all observing females. Topi and lechwes antelope are often seen using this strategy in areas where it is worthwhile for males to defend territory, such as those with optimal resources and low predation. When this is not the case, and the cost of maintaining territorial boundaries outweighs the benefits, females and males will abandon this strategy. Males will cease fighting for territory and will mate with any available females; this strategy comes with its own consequences, as harassing males may interrupting mating episodes.

Polygamous Males

Some larger male antelope will graze across larger distances, competing with males for herds of females. Once a male wins a dominant female over, her herd will become stationary and he can then potentially mate with all members. As calving season comes along, these groups tend to splinter, forming smaller, segregated groups, perhaps to protect themselves from predation. Bachelor males move on, repeating the process the following year.

The Cowardly Antelope

Recent studies looking more closely into antelope mating behavior, especially lekking, have found an interesting fact. You might imagine males battling over females would want to come across as strong and confident. However, some males, fearing they will not be able to mate, will as a female walks away let out a vocal signal indicating predator danger when no such danger exists. This causes a female to pause, and gives the male another opportunity to win her over.

CONSERVATION

Living in a densely populated part of the world, the four-horned antelope is threatened by loss of its natural habitat to agricultural land. In addition, the unusual four-horned skull has been a popular target for trophy hunters. Only around 10,000 four-horned antelope are estimated to remain alive in the wild, although many are in protected animal conservation areas. The species is protected under the Indian Wildlife Protection Act and the Nepalese population. The four-horned antelope is considered Vulnerable by the IUCN, primarily due to increasing habitat loss.

CHAPTER 8
INDIAN RHINOCEROS

CLASSIFICATION
Kingdom: Animalia, Phylum: Chordata, Class: Mammalia, Order: Rhinocerotidae, Family: Perissodactyla, Genus: *Rhinoceros,* Species: *unicornis*

GENERAL CHARACTERS
The **Indian rhinoceros** (*Rhinoceros unicornis*), also called the **greater one-horned rhinoceros** and **Indian one-horned rhinoceros**, belongs to the family Rhinocerotidae. Listed as a **vulnerable** species, this large mammal is primarily found in India's Assam, West Bengal and in protected areas in the Terai of Nepal, where populations are confined to the riverine grasslands in the foothills of the Himalayas. The Indian rhinoceros once ranged throughout the entire stretch of the Indo-Gangetic Plain, but excessive hunting reduced their range drastically. Today, more than 3,000 rhinos live in the wild. In 2014, 2,544 of which are found in India's Assam alone, an increase by 27 percent since 2006, although in early the 1900s, Assam had about 200 rhinos only. It is the fifth largest land animal. The Indian rhinoceros has a thick grey-brown skin with pinkish skin folds and a black horn. Its upper legs and shoulders are covered in wart-like bumps. It has very little body hair, aside from eyelashes, ear fringes and tail brush. Males have huge neck folds. Its skull is heavy with a basal length above 60 cm (24 in) and an occiput above 19 cm (7.5 in). Its nasal horn is slightly back-curved with a base of about 18.5 cm (7.3 in) by 12 cm (4.7 in) that rapidly narrows until a smooth, even stem part begins about 55 mm (2.2 in) above base. In captive animals, the horn is frequently worn down to a thick knob. The rhino's single horn is present in both males and females, but not on newborn young. The black horn is pure keratin, like human fingernails, and starts to

show after about six years. In most adults, the horn reaches a length of about 25 cm (9.8 in), but has been recorded up to 57.2 cm (22.5 in) in length.Among terrestrial land mammals native to Asia, the Indian rhinoceros is second in size only to the Asian elephant. It is also the second-largest living rhinoceros, behind only the white rhinoceros. Males have an average head and body lengthof 368–380 cm (12.07–12.47 ft) with a shoulder height of 170–186 cm (5.58–6.10 ft), while females have an average head and body length of 310–340 cm (10.2–11.2 ft) and a shoulder height of 148–173 cm (4.86–5.68 ft). The male, averaging 2,200 kg (4,900 lb) is heavier than the female, at an average of 1,600 kg (3,500 lb).The rich presence of blood vessels underneath the tissues in folds gives it the pinkish colour. The folds in the skin increase the surface area and help in regulating the body temperature. The thick skin does not protect against blood sucking *Tabanus* flies, leeches and ticks. The largest sized specimens range up to 4,000 kg (8,800 lb).

HABIT AND HABITAT

Indian rhinoceros are **grazers**. Their diets consist almost entirely of grasses, but they also eat leaves, branches of shrubs and trees, fruits, and submerged and floating **aquatic plants**. They feed in the mornings and evenings. They use their prehensile lips to grasp grass stems, bend the stem down, bite off the top, and then eat the grass. They tackle very tall grasses or saplings by walking over the plant, with legs on both sides and using the weight of their bodies to push the end of the plant down to the level of the mouth. Mothers also use this technique to make food edible for their calves. They drink for a minute or two at a time, often imbibing water filled with rhinoceros urine. One-horned rhinos once ranged across the entire northern part of the Indian Subcontinent, along the Indus, Ganges and Brahmaputra River basins, from Pakistan to the Indian-Burmese border, including Bangladesh and the southern parts of Nepal and Bhutan. They may have also occurred in Myanmar, southern China and Indochina. They inhabit the alluvial plain grasslands of the Terai and the Brahmaputra basin. As a result of habitat destruction and climatic changes their range has gradually been reduced so that by the 19th century, they only survived in the Terai grasslands of southern Nepal, northern Uttar Pradesh, northern Bihar, northern Bengal, and in the Brahmaputra Valley of Assam. One-horned rhinos once ranged across the entire northern part of the Indian Subcontinent, along the Indus, Ganges and Brahmaputra River basins, from Pakistan to the Indian-Burmese border, including Bangladesh and the southern parts of Nepal and Bhutan. They may have also occurred in Myanmar, southern China and Indochina. They inhabit the alluvial plain grasslands of the Terai and the Brahmaputra basin. As a result of habitat destruction and climatic changes their range has gradually been reduced so that by the 19th century, they only survived in the Terai grasslands of southern Nepal, northern Uttar Pradesh, northern Bihar, northern Bengal, and in the Brahmaputra Valley of Assam.

POPULATIONS

In 2007, the total population was estimated to be 2,575 individuals, of which 2,200 lived in Indian protected areas:

- in Kaziranga National Park: 2329 (2012 estimate)— increased from 366 in 1966
- in Jaldapara National Park: 108 — increased from 84 in 2002
- in Pobitora Wildlife Sanctuary: 81 — increased from 54 in 1987
- in Orang National Park: 68 — increased from 35 in 1972
- in Gorumara: 27 — increased from 22 in 2002
- in Dudhwa National Park: 21
- in Manas National Park: 19
- in Katarniaghat Wildlife Sanctuary: 2

Pobitora Wildlife Sanctuary shelters the highest density of Indian rhinos in the world — with 84 individuals in 2009 in an area of 38.80 km^2 (14.98 sq mi).The population in Nepal increased by 111 individuals from 2011 to 2015, increasing by 21%. The latest rhino count was conducted from 11 April to 2 May 2015 and revealed 645 individuals living in Parsa Wildlife Reserve, Chitwan National Park, Bardia National Park, Shuklaphanta Wildlife Reserve and respective buffer zones in the Terai Arc Landscape. In Pakistan's LalSuhanra National Park, two rhinos from Nepal were introduced in 1983 but have not bred so far.

ECOLOGY AND BEHAVIOUR

Rhinos are mostly **solitary** creatures, with the exception of mothers and calves and breeding pairs, although they sometimes congregate at bathing areas. They have home ranges, those of males being usually 2 to 8 km^2 (0.77 to 3.09 sq mi) and overlapping each other. Dominant males tolerate males passing through their territories except when they are in mating season, when dangerous fights break out. They are active at night and early morning. They spend the middle of the day wallowing in lakes, rivers, ponds, and puddles to cool down. They are very good swimmers. Over 10 distinct vocalizations have been recorded. Indian rhinos bathe regularly. The folds in their skin trap water and hold it even when they come back on land. Indian rhinos have few natural enemies, except for tigers, which sometimes kill unguarded calves, but adult rhinos are less vulnerable due to their size. Mynahs and egrets both eat invertebrates from the rhino's skin and around its feet. *Tabanus* flies, a type of horse-fly, are known to bite rhinos. The rhinos are also vulnerable to diseases spread by parasites such as leeches, ticks, and nematodes. Anthrax and the blood-disease septicemia are known to occur. They can run at speeds of up to 55 km/h (34 mph) for short periods and are excellent swimmers. They have excellent senses of hearing and smell, but relatively poor eyesight.

SOCIAL LIFE

The Indian rhinoceros forms a variety of social groupings. Adult males are generally solitary, except for mating and fighting. Adult females are largely solitary when they are without calves. Mothers will stay close to their calves for up to four years after their birth, sometimes allowing an older calf to continue to accompany her once a newborn calf arrives. Sub adult males and females form consistent groupings, as well. Groups of two or three young males will often form on the edge of the home ranges of dominant males, presumably for protection in numbers. Young females are slightly less social than the males. Indian rhinos also form short-term groupings, particularly at forest wallows during the monsoon season and in grasslands during March and April. Groups of up to 10 rhinos may gather in wallows—typically a dominant male with females and calves, but no sub adult males.

The Indian rhinoceros makes a wide variety of vocalizations. At least 10 distinct vocalizations have been identified: snorting, honking, bleating, roaring, squeak-panting, moo-grunting, shrieking, groaning, rumbling and humphing. In addition to noises, the rhino uses olfactory communication. Adult males urinate backwards, as far as 3–4 m behind them, often in response to being disturbed by observers. Like all rhinos, the Indian rhinoceros often defecates near other large dung piles. The Indian rhino has pedal scent glands which are used to mark their presence at these rhino latrines. Males have been observed walking with their heads to the ground as if sniffing, presumably following the scent of females.

In aggregations, Indian rhinos are often friendly. They will often greet each other by waving or bobbing their heads, mounting flanks, nuzzling noses, or licking. Rhinos will playfully spar, run around, and play with twigs in their mouths. Adult males are the primary instigators in fights. Fights between dominant males are the most common cause of rhino mortality, and males are also very aggressive toward females during courtship. Males will chase females over long distances and even attack them face-to-face. Unlike African rhinos, the Indian rhino fights with its incisors, rather than its horns.

REPRODUCTION

Captive males breed at five years of age, but wild males attain dominance much later when they are larger. In one five-year field study, only one rhino estimated to be younger than 15 years mated successfully. Captive females breed as young as four years of age, but in the wild, they usually start breeding only when six years old, which likely indicates they need to be large enough to avoid being killed by aggressive males. Their gestation period is around 15.7 months, and birth interval ranges from 34–51 months. In captivity, four rhinos are known to have lived over 40 years, the oldest living to be 47.

CONSERVATION

The Indian and Nepalese governments have taken major steps towards Indian rhinoceros conservation, especially with the help of the World Wide Fund for Nature (WWF) and other nongovernmental organizations. In the early 1980s, a rhino translocation scheme was initiated. The first pair of rhinos was reintroduced from Nepal's Terai to Pakistan's LalSuhanra National Park in Punjab in 1982. In 1910, all rhino hunting in India became prohibited. In 1984, five rhinos were relocated to Dudhwa National Park — four from the fields outside the Pabitora Wildlife Sanctuary and one from Goalpara. In 1957, the country's first conservation law insured the protection of rhinos and their habitat. In 1959, Edward Pritchard Gee undertook a survey of the Chitwan Valley, and recommended the creation of a protected area north of the Rapti River and of a wildlife sanctuary south of the river for a trial period of 10 years. After his subsequent survey of Chitwan in 1963, he recommended extension of the sanctuary to the south. By the end of the 1960s, only 95 rhinos remained in the Chitwan Valley. The dramatic decline of the rhino population and the extent of poaching prompted the government to institute the *GaidaGasti* – a rhino reconnaissance patrol of 130 armed men and a network of guard posts all over Chitwan. To prevent the extinction of rhinos, the Chitwan National Park was gazetted in December 1970, with borders delineated thefollowing year and established in 1973, initially encompassing an area of 544 km^2 (210 sq mi). Since 1973, the population has recovered well and increased to 645 animals in 2015.To ensure the survival of rhinos in case of epidemics, animals were translocated annually from Chitwan to the Bardia National Park and the Sukla Phanta Wildlife Reserve since 1986.

CHAPTER 9
IRRAWADDY DOLPHIN (SNUBFIN DOLPHIN)

CLASSIFICATION
Kingdom: Animalia, Phylum: Chordata, Class: Mammalia, Order: Cetacea, Family: Delphinidae, Genus: *Orcaella*, Species: *brevirostris*

GENERAL CHARACTERS
The **Irrawaddy dolphin** (*Orcaella brevirostris*) is a marine mammal in the family of oceanic dolphins, part of the order of cetaceans. A particularly distinctive dolphin, the Irrawaddy has a rounded head with no beak and a flexible neck, causing visible creases behind the head. Although most closely related to the orca, the Irrawaddy dolphin is similar in body form to the Beluga whale, but darker in colour, with a pale to dark grey back and a light underside. The dorsal fin is small, triangular and rounded, and the flippers are long and broad. The color of the Irrawaddy dolphin has been described as dark bluish grey, slate grey, battleship grey, or pale grey. The underside is usually paler than the dorsal surface. *Orcaella brevirostris* has a high, anteriorly convex forehead which overhangs the mouth. It does not have a beak, and its U-shaped blowhole is to the left of the midline. Unlike the condition in most dolphin species, the blowhole opens toward the front. Irrawaddy dolphins resemble the finless porpoise, but unlike that species, they have a small, triangular, and bluntly rounded dorsal fin (with a barely concave rear margin) set just behind the midback. The flippers are relatively large (about one-sixth as long as the body) and have a great breadth with a gently curved leading edge. The mouth line is straight, and there may be a visible neck crease. The neck is unusually flexible because only the first two cervical vertebrae are fused. The tail is also quite flexible. *Orcaella brevirostris* has homodont, narrow, pointed, and peg-like

teeth with slightly expanded crowns. The teeth are about 1 cm in length, and tooth counts are 17 to 20 (upper) and 15 to 18 (lower) ineach quadrant. The skull is characterized by its globular shape, short rostrum, and broad facial region. The Irrawaddy dolphin does not have a cardiac sphincter; the stomach is subdivided into compartments communicating through narrow orifices.

HABIT AND HABITAT

Orcaella brevirostris is a generalist feeder, taking food both from within the water column and from the bottom. Bony fishes seem to be the main food of Irrawaddy dolphins, but they have also been observed to eat crustraceans, cephalopods, and fish eggs. Stomachs of ten *Orcaella brevirostris* from coastal waters off Townsville, Australia all contained bony fishes (from 16 orders and 13 families). Nine of the ten stomachs contained crustaceans (five with shrimps, two with isopods, and four with unidentified crustaceans), and all stomachs contained cephalopod remains (ten with squid, three with cuttlefish, and two with octopods). Two species of cyprinid fish, *Cirrihinus siamensis* and *Paralaubuca typus*, are believed to be an especially important food source for Irrawaddy dolphins in northeastern Cambodia and Lao PDR. Carp appear to be the most important food source for Irrawaddy dolphins in Semanyang Lake (Kalimantan).

Irrawaddy dolphins inhabit coastal, brackish, and fresh waters (major river systems) of the tropical and sub-tropical Indo-Pacific. They have been found as far as 1440 km upstream and can live permanently in freshwater. The Irrawaddy dolphin has a patchy distribution in the shallow, coastal waters of the Indo-Pacific from northern Australia and the Philippines to northeastern India. There are a number of sub poulations, including:

- Ayeyarwady River subpopulation of Burma
- Mahakam River subpopulation of Indonesia
- Mekong River subpopulation Laos, Cambodia and Viet Nam
- Malampaya Sound subpopulation in Palawan, Philippines
- Songkhla Lake subpopulation in Thailand
- Chilka Lake subpopulation in India

SPECIAL CHARACTERS

Irrawaddy dolphins are not notably active, but they do make low leaps on occasion. They are usually seen while surfacing slowly and exposing their blowhole. The animal then usually continues forward with a smooth slow roll. They are not known to bowride. Nothing is known of the depths to which Irrawaddy dolphins dive, but considering their coastal and riverine distribution, it is unlikely for them to dive to considerable depths. It has been reported that *Orcaella brevirostris* typically respires three times in rapid succession and then dives for 30-60 seconds (dive times are longer when the animal is frightened). The maximum dive time recorded is 12 minutes, and the maximum swimming rate recorded is 25 km/hr (recorded while a *Orcaella brevirostris* was being chased by a boat). Irrawaddy dolphins are usually seen in small groups, which usually consist of less than six individuals but which may contain as many as 10 to 15 animals. There is no information on the population dynamics of this species. The behavior of *Orcaella brevirostris* suggests that they spend most of their time feeding. Irrawaddy dolphins sometimes spit water while feeding (they can expel water from their mouths for distances of up to 1.5 m), apparently to herd fish. Fishermen also allege that the dolphins sometimes catch large fish for sport by stunning them with a blow from their lower jaw. The dolphins then play with the fish like a cat with a mouse before discarding them.The vocalizations of Irrawaddy dolphins are short time-duration band with signals of about 25-30 microsecond duration. The main sonar signal consists of only a few cycles of a dominant frequency of around 60 Kilohertz. Pulse trains are rather regular in nature. No audible whistles or pure tones have been recorded. Believed to be reincarnated humans by some of the people of Laos, Irrawaddy dolphins are less active than many other dolphins, making only occasional low leaps and never bow-riding. Feeding together in groups of usually less than six, but as many as 15, the Irrawaddy dolphin can dive for up to 12 minutes to feed on bony fish, crustaceans, cephalopods and fish eggs. Irrawaddy dolphins are known to spit water to herd fish, and have even been reported to stun large fish with a blow from the lower jaw, only to play with them before casting them aside. In some areas of Asia, fishermen consider the Irrawaddy dolphin to be a competitor for fish, but in other areas the fishermen attract them to the boat and encourage them to drive fish into the nets for a share of the catch. Irrawaddy dolphins communicate with clicks, creaks and buzzes. This dolphin species is known to carry out daily migrations from the Semayang Lake in eastern Borneo to the Mahakam River, returning to the lake in the evening. In Indonesia, Irrawaddy dolphins move into tributaries at high water and into the main river during low water. The individuals found in northern Australia, which are morphologically distinct from Asian individuals, do not appear to migrate. Reports from various parts of Asia suggest that Irrawaddy dolphins regularly assist fisherman by driving fish into their nets. In one report, fisherman in Burma were observed to attract Irrawaddy dolphins by tapping the sides of their boats with oars. The dolphins swim around the boats in ever-diminishing circles, thereby forcing fish into nets. The fisherman share their catch with the dolphins and consider them friends that are not to be harmed.

REPRODUCTION
Little is known about the reproduction/breeding behaviour of Irrawaddy dolphins, but according to Kampuchean fisherman, the courtship season is from March to June at 11-12 degrees north latitude. These fisherman say they observe copulation almost daily during this period, and fights between males are also often observed. Coitus is preceded by much play including chases or jumps with the partners often leaping out of the water, belly to belly. During coitus, other members of the group swim around the mating pair. The Irrawaddy dolphin has not been extensively studied, and little is known about its reproduction and breeding behavior. The mating season is believed to extend from April to June in the Semayang Lake/Mahakam River area of Kalimantan. Calves from animals caught in this area have been born in captivity in Jakarta (six degrees north of Kalimantan) in July and December. The age of sexual maturity is unknown, but there is evidence that at least some dolphins reach adult size when they are four to six years old. The gestation period is estimated to be fourteen months. A neonate born in captivity in Jakarta was 96 cm long and weighed 12.3 kg. It was born twelve days after milk was first seen discharging from its mother, and the tail was observed protruding from the genital slit more than two hours before the calf was born. The calf started suckling twelve hours after birth and eating dead fishes at the age of six months. It was fully weaned by two years of age. During its first seven months, the calf increased in length by 57 cm (59%) and in weight by 32.7 kg (266%). Little is known about the reproductive biology of the Irrawaddy dolphin, but it is thought to breed between April and June in the Mahakam River, and gestation is estimated at 14 months. The species lifespan in the wild is estimated to be approximately 25 years.

CONSERVATION STATUS
All cetaceans are protected under Australian legislation within the Australian Exclusive Economic zone which extends 200 nm off the northern coast. The Irrawaddy dolphin is also protected by law in Laos. Some captive breeding of this species has been successful. They are classified in the IUCN Red list as vulnerable.

Threats
It has been estimated that there are fewer than 2,000 Irrawaddy dolphins left, but they are not believed to be at risk of imminent extinction. Most live captures are for the oceanarium trade in Asia, and hunting of this species is rare, occuring only in parts of India to harvest oil for the treatment of rheumatism. However, the Irrawaddy dolphin is likely to be affected by increasing pollution, construction of dams and the build-up of silt following severe erosion. Incidental catches occur and fishing with explosives also results in dolphin casualties. Also, shark gillnets in Australia

and fish traps and other types of nets throughout the range are known to take away some Irrawaddy dolphins. Furthermore, some small-scale hunting by local people probably occurs in many areas of its range (nonetheless, the Irrawaddy dolphin is generally unexploited), and the Irrawaddy dolphin inhabits some of the most vulnerable of aquatic habitats, those being tropical, riverine, estuarine, and coastal habitats. The oil of the Irrawaddy dolphin has reportedly been used as a remedy for rheumatism in parts of India.

CHAPTER 10
INDIAN PEAFOWL

Male (peacock)　　　　　　Female (peahen)

CLASSIFICATION
Kingdom: **Animalia,** Phylum: **Chordata,** Class: **Aves,** Order: **Galliformes,**
Family: **Phasianidae,** Subfamily: **Phasianinae,** Genus: ***Pavo,*** Species: *cristatus*

GENERAL CHARACTERS
The **Indian peafowl** or **blue peafowl** (*Pavo cristatus*), a large and brightly coloured bird, is a species of peafowl native to South Asia, but introduced in many other parts of the world like the United States, Mexico, Honduras, Colombia, Guyana, Suriname, Brazil, Uruguay, Argentina, South Africa, Madagascar, Mauritius, Réunion, Indonesia, Papua New Guinea and Australia. The species was first named and described by Linnaeus in 1758, and the name *Pavo cristatus* is still in use now. The male peacock is predominantly blue with a fan-like crest of spatula-tipped wire-like feathers and is best known for the long train made up of elongated upper-tail covert feathers which bear colourful eyespots. These stiff feathers are raised into a fan and quivered in a display during courtship. Females lack the train, and have a greenish lower neck and duller brown plumage. The Indian peafowl lives mainly on the ground in open forest or on land under cultivation where they forage for berries, grains but also prey on snakes, lizards, and small rodents. Their loud calls make them easy to detect, and in forest areas often indicate the presence of a predator such as a tiger. They forage on the ground in small groups and usually try to escape on foot through undergrowth and avoid flying, though they fly into tall trees to roost.

The function of the peacock's elaborate train has been debated for over a century. In the 19th century, Charles Darwin found it a puzzle, hard to explain through ordinarynatural selection. His later explanation, sexual selection, is widely but not universally accepted. In the 20th century, AmotzZahavi argued that the train was a handicap, and that males were honestly signalling their fitness in proportion to the splendour of their trains. Despite extensive study, opinions remain divided on the mechanisms involved.

The bird is celebrated in Indian and Greek mythology and is the national bird of India. The Indian peafowl is listed as of **Least Concern** by the International Union for Conservation of Nature (IUCN).Peacocks are a larger sized bird with a length from bill to tail of 100 to 115 cm (40 to 46 inches) and to the end of a fully grown train as much as 195 to 225 cm (78 to 90 inches) and weigh 4–6 kg (8.8–13.2 lbs). The females, or peahens, are smaller at around 95 cm (38 inches) in length and weigh 2.75–4 kg (6–8.8 lbs). Indian peafowl are among the largest and heaviest representatives of the **Phasianidae**. Their size, colour and shape of crest make them unmistakable within their native distribution range. The male is metallic blue on the crown, the feathers of the head being short and curled. The fan-shaped crest on the head is made of feathers with bare black shafts and tipped with blush-green webbing. A white stripe above the eye and a crescent shaped white patch below the eye are formed by bare white skin. The sides of the head have iridescent greenish blue feathers. The back has scaly bronze-green feathers with black and copper markings. The scapular and the wings are buff and barred in black, the primaries are chestnut and the secondaries are black. The tail is dark brown and the "train" is made up of elongated upper tail coverts (more than 200 feathers, the actual tail has only 20 feathers) and nearly all of these feathers end with an elaborate eye-spot. A few of the outer feathers lack the spot and end in a crescent shaped black tip. The underside is dark glossy green shading into blackish under the tail. The thighs are buff coloured. The male has a spur on the leg above the hind toe.

HABIT AND HABITAT

Peafowl are omnivorous and eat seeds, insects, fruits, small mammals and reptiles. They feed on small snakes but keep their distance from larger ones. In the Gir forest of Gujarat, a large percentage of their food is made up of the fallen berries of *Zizyphus*. Around cultivated areas, peafowl feed on a wide range of crops such as groundnut, tomato, paddy, chilly and even bananas. Around human habitations, they feed on a variety of food scraps and even human excreta. In the countryside, it is particularly partial to crops and garden plants.The Indian peafowl is a resident breeder across the Indian subcontinent and is found in the drier lowland areas of Sri Lanka. In South Asia, it is found mainly below an altitude of 1,800 metres (1.1 mi) and in rare cases seen at about 2,000 metres (1.2 mi). It is found in moist and dry-deciduous forests, but can adapt to live in cultivated regions and around human habitations and is usually found where water is available. In many parts of northern India, they are protected by religious practices and will forage around villages and towns for scraps. Some have suggested that the peacock was introduced into Europe by Alexander the Great, while others say the bird had reached Athens by 450 BC and may have been introduced even earlier. It has since been introduced in many other parts of the world and has become feral in some areas. In isolated cases, the Indian peafowl has been known to be able to adapt to harsher climates, such as those of northern Canada. The species has been spotted as far north as Schomberg, Ontario, thriving in its newly adapted northern climate.

REPRODUCTION

Peacocks are polygamous, and the breeding season is spread out but appears to be dependent on the rains. Peafowls usually reach sexual maturity at the age of 2 to 3 years old. Several males may congregate at a lek site and these males are often closely related. Males at lek appear to maintain small territories next to each other and they allow females to visit them and make no attempt to guard harems. Females do not appear to favour specific males. The males display in courtship by raising the upper-tail coverts into an arched fan. The wings are held half open and drooped and it periodically vibrates the long feathers producing a ruffling sound. The cock faces the hen initially and struts and prances around and sometimes turns around to display the tail. Males may also freeze over food to invite a female in a form of courtship feeding. Males may display even in the absence of females. When a male is displaying, females do not appear to show any interest and usually continue their foraging. The peak season in southern India is April to May, January to March in Sri Lanka and June in northern India. The nest is a shallow scrape in the ground lined with leaves, sticks and other debris. Nests are sometimes placed on buildings and in earlier times have been recorded using the disused nest platforms of the white-rumped vultures. The clutch consists of 4–8 fawn to buff white eggs which are incubated only by the female. The eggs take about 28 days to hatch. The chicks are nidifugous and follow the mother around after hatching. Downy young may sometimes climb on their mothers' back and the female may carry them in flight to a safe tree branch. An unusual instance of a male incubating a clutch of eggs has been reported. In captivity, birds have been known to live for 23 years but it is estimated that they live for only about 15 years in the wild.

CHAPTER 11
GHARIAL

CLASSIFICATION
Kingdom: Animalia, Phylum: Chordata, Class: Reptilia, Order: Crocodilia, Family: Gavialidae, Genus: *Gavialis,* Species: *gangeticus*

GENERAL CHARACTERS

The **Gharial** (*Gavialis gangeticus*) is one of two surviving members of the family Gavialidae, a long-established group of crocodile-like reptiles with long, narrow jaws. The Gharial (sometimes called the 'Indian gharial' or 'gavial') is the second-longest of all living crocodilians. Gharials are most adapted to the calmer areas in the deep fast moving rivers. The physical attributes of the gharial do not make it very suited for moving about on land. In fact the only reasons the gharial leaves the water is to either bask in the sun or to nest on the sandbanks of the rivers. Gharials have elongated, narrow snouts that are similar only to its relative, the Falsegharial, (Tomistomaschlegelii). The snout shape varies with the age of the Gharial. The snout becomes progressively thinner the older the gharial gets. The bulbous growth on the tip of the males snout is called a 'ghara' (after the Indian word meaning 'pot'), only present in mature individuals.

The bulbous growth is used for various activities, it is used to generate an echoing 'hum' during vocalization, it acts as a visual lure for attracting females and it is also used to make bubbles which have been associated with the mating rituals of the species. Gharials elongated jaws are lined with many interlocking, razor-sharp teeth, an adaptation to the diet (predominantly fish in adults). Being one of the largest of all crocodilian species, approaching the size of the Salt water crocodile (*Crocodylus porosus*) and the Nile crocodile in maximum size, the males reach at least 5 – 6 metres in length. Reports of 7 metre gharials exist, but are unconfirmed.The leg muscular system of the gharial is not suited to enable the animal to raise the body off the ground (on land) in order to achieve the high-walk gait, being able only to push its body forward across the ground in a movement called 'body sliding'. Although the gharial can do this with some speed when required, when in water, the gharial is the most nimble and quickest of all the crocodiles in the world. Their tail seems overdeveloped and is laterally flattened, more so than other crocodiles, this enables it to achieve the excellent water locomotive abilities.

HABIT AND HABITAT

The diet of juvenile gharials is different from adults. Juveniles eat small animals, such as insects, crustaceans, or frogs. But as they grow older and their snout becomes thinner and longer, they eat almost exclusively fish. Their jaws are well adapted for catching fish. There are three main hunting strategies. The sit and wait approach is where they float almost completely submerged under water and remain motionless until their pray passes right by them. The sweeping search involves an integumentary sensory organ found on the scales to sense vibrations in the water while slowly feeling through the water for prey. The third hunting strategy is a rapid strike. The thin jaw creates low water resistance for quick snaps underwater. Indian gharials live in clear freshwater rivers with fast flowing currents. They congregate at river bends and other sections of rivers where the water is deep and the current is reduced. Because Indian gharials are not well adapted for movement on land, they usually leave the water only to bask and to nest. They prefer sandbars in the middle of the rivers for both of these activities. Juveniles may seek out quiet backwaters or smaller streams.

SPECIAL CHARACTERS

Indian gharials spend a lot of time basking in the sun, more so in the winter than in the summer. They tend to revisit the same basking spot, which is always close to the water. Indian gharials also "gape" during basking to dissipate excess heat. Gaping is usually done in 10 to 20 minute intervals with the head at a 20 degree angle. On very hot days gharials completely submerge their bodies, leaving only their heads out of the water at a 20 to 30 degree angle. Indian gharials aggregate in basking and nesting areas but are generally solitary. Nests are defended by females.

Like all crocodilians, Indian gharials possess integumentary sense organs. These are tiny pits in the scales that cover the body. These pits are able to pick up vibrations or changes in water pressure, which aid in the search for prey. Their eyes have a reflective layer behind the eye, the tapetumlucidum, which aids in night vision. A clear membrane, the nictitating membrane, protects the eye while under water. Indian gharials pick up low frequencies through hearing and are able to close the ear canal when submerged. Indian gharials apparently communicate via vibrations in the water and buzzing sounds made by males with the ghara on their snouts.

REPRODUCTION

The "gharal" is used in mating. This is a cartilaginous lid on the nostril of males that flaps when exhaling, producing a loud buzzing noise, which is used during territorial defense and courtship. Males also hiss, and perform above water jaw slapping. While underwater, jaw slapping is also performed to attract possible mates. When a female finds a male, they will rub each other with their snouts and the male will follow the female around his territory. The female will then show her readiness to mate by raising her head skyward, at which point the male will climb on top of her. The two will then submerge for up to 30 minutes during copulation.

Mating season occurs for about two months each year. Mating type is polygynous. Mating season varies regionally, but generally occurs between November and February, during the dry season. Nesting occurs during the late dry season, from March through May. Females locate a steep sand bank where they dig a nest. During this time, they might dig a number of holes before finding the right spot. Holes are about 50 cm deep and from 3 to 5 meters from the water. Females lay 28 to 60 eggs in the hole, usually at night. Very large females are capable of laying almost 100 eggs. An average Indian gharial egg is 5.5 centimeters wide, 8.6 centimeters long, and weighs 100 to 156 grams. An incubation period of 60 to 80 days will follow. Females continue to visit and guard eggs during the night but remain in the water during the day. During incubation females are very territorial near the nest, but they tolerate other females nesting on the same beach. Nests in warmer climates usually hatch earlier. Young are about 18 cm in length. The female (and perhaps the male) will help excavate the nest during hatching, but they are probably incapable of picking up the young. Sexual maturity for females is reached at 8 years old and 3 meters in length. For males, maturity is attained at 15 years of age and 4 meters in length. At this time males grow a ghara on their snout. Females must provision eggs with yolk prior to oviposition, excavate a nest cavity, and guard nests. Females may uncover and assist young during the hatching process. After hatching, females protect hatchlings for several weeks, often until monsoon rains come, during which high water levels may disperse the young. The male will be tolerated nearby, but they do not actively protect hatchlings, though young will sometimes rest on the back of the male. The only record of longevity in *Gavialis gangeticus* is of a captive individual at the London Zoo, where one was estimated at 29 years old. Because of their large size, it is thought that they have a long life span. Fisherman that live near gharials believe that they can live as long as 100 years old, though this has not been confirmed.

Economic Importance for Humans

Male Indian gharials are sometimes sought after for their ghara, the growth on the end of their snout, because it is believed by some to carry aphrodisiac properties. Eggs are collected for their supposed medicinal properties. However, both of these supposed medicinal properties are not based on research and it is unlikely that the eggs or the ghara benefit people in any way. Indian gharials may benefit local communities by acting as a tourist attraction. This species is probably harmless to human interests. Indian gharials are sometimes believed to attack and eat humans, but this appears to be an unfounded fear. Indian gharials are generally not aggressive and have narrow jaws and thin teeth that are unsuited to attacking humans or large animals.

CONSERVATION STATUS

The decline from an estimated 436 adult Indian gharials in 1997 to 182 in 2006 represents a 58% drop across their range. This drastic decline happened in a period of nine years, well within the span of one generation, qualifying Indian gharials as Critically Endangered (IUCN). They were the first crocodilian to be categorized as critically endangered. The biggest threat to them is habitat loss and disturbance caused by people clearing riparian areas for firewood or farmland or mining river banks for sand. Poaching is also a problem. Conservation efforts have increased in recent years and attempts to ensure population increases are in place. Action groups such as the Gharial Multi-Task Force are comprised of regional and international crocodilian specialists that are working to avoid the extinction of this animal in the wild. Information about the current status of Indian gharials in the wild is still being collected. Conservation efforts andmanagement strategies cannot be put into place without good data to back them. Surveys of areas such as Pakistan and Burma are some of the next steps to be taken. Other threats to this species includea lack of proper release sites. Eggs are collected by local people for medicinal purposes and adult males are hunted because it is believed that the ghara on their snout acts as an aphrodisiac. Fishing also causes a problem when they are captured by gill nets and are killed in the process. Fishing also greatly reduces the prey base of these animals. It is thought by some local people that Indian gharials are man-eaters, which results in persecution. This fear stems from the fact that human remains are sometimes found in the bellies of gharials. During a Hindu funeral ritual, cremated remains of a body are placed in rivers. It is a common practice for many crocodilians to ingest rocks to be used as gastroliths: hard objects that aid in digestion and alter their buoyancy. It is thought that some human remains and jewelry is ingested in a similar way. Indian gharial jaws are specialized for eating fish and they are not considered dangerous to people. A recent threat to the species is a widespread mortality due to gout. Since 2007 over 110 gharials have succumbed to gout. This may be caused by the introduction of *Tilapia* into the Yamuna river. It is believed that these fish carry a toxin that effects gharials, but the composition of the toxin and how it enters the river is still being researched.

CHAPTER 12
SEA TURTLES

CLASSIFICATION
Kingdom: animalia, Phylum: Chordata, Class: Reptilia, Order: Testudines, Family: Cheloniidae, Genus: *Chelonia*, Species: *mydas*

GENERAL CHARACTERS
Sea turtles (super family **Chelonioidea**), sometimes called **marine turtles**, are reptiles of the order Testudines. There are seven species of sea turtles. They are the leatherback sea turtle, green sea turtle, loggerhead sea turtle, Kemp's ridley sea turtle, hawksbill sea turtle, flatback sea turtle and olive ridley sea turtle. Four of the species have been identified as "endangered" or "critically endangered" with another two being classed as "vulnerable.

The seven living species of sea turtles are: leatherback sea turtle, green sea turtle, loggerhead sea turtle, Kemp's ridley sea turtle, hawksbill sea turtle, flatback sea turtle and olive ridley sea turtle. All species except the leatherback are in the family Cheloniidae. The leatherback belongs to the family Dermochelyidae and is its only member.

The species are primarily distinguished by their anatomy: for instance, the prefrontal scales on the head, the number of and shape of scutes on the carapace, and the type of inframarginal scutes on the plastron. The leatherback is the only sea turtle that does not have a hard shell; instead, it bears a mosaic of bony plates beneath its leathery skin. It is the largest sea turtle, measuring 6 to 9 feet (1.8 to 2.7 m) in length at maturity, and 3 to 5 feet (0.91 to 1.52 m) in width, weighing up to 1,500 pounds (680 kg). Other species are smaller, being mostly 2 to 4 feet (0.61 to 1.22 m) and proportionally narrower.

HABIT AND HABITAT

Depending on the species, sea turtles may be **carnivorous** (meat eating), **herbivorous** (plant eating), or **omnivorous** (eating both meat and plants). The jaw structure of many species is adapted for their diet.

Green sea turtles have finely serrated jaws adapted for a vegetarian diet of sea grasses and algae. As adults, these are the only herbivorous sea turtles. Some species change eating habits as they age. For example, green sea turtles are mainly carnivorous from hatching until juvenile size; they then progressively shift to an herbivorous diet. A hawksbill has a narrow head with jaws meeting at an acute angle, adapted for getting food from crevices in coral reefs. They eat sponges, tunicates, shrimps, and squids. Loggerheads' and ridleys' jaws are adapted for crushing and grinding. Their diet consists primarily of crabs, molluscs, shrimps, jellyfish, and vegetation. Leatherbacks have delicate scissor-like jaws that would be damaged by anything other than their normal diet of jellyfish, tunicates, and other soft-bodied animals. The mouth cavity and throat are lined with papillae (spine-like projections) pointed backward to help them swallow soft foods. Researchers continue to study the feeding habits of flatbacks. There is evidence that they are opportunistic feeders that eat seaweeds, cuttlefish, and sea cucumbers. In a zoological environment all sea turtle species can be maintained on a carnivorous diet. Adults of most species are found in shallow, coastal waters, bays, lagoons, and estuaries. Some also venture into the open sea. Juveniles of some species may be found in bays and estuaries, as well as at sea.

BEHAVIOUR

Social Behaviour

Sea turtles are not generally considered social animals; however, some species do congregate offshore.

Sea turtles gather together to mate. Members of some species travel together to nesting grounds. After hatchlings reach the water they generally remain solitary until they mate.

Individual Behaviour

Little is known about the individual behavior of sea turtle species. In the ocean, flatback turtles may spend hours at the surface floating, apparently asleep or basking in the sun. Frequently, seabirds perch on the backs of the flatbacks. Hawksbill turtles spend some time resting or sleeping wedged into coral or rock ledges. Olive ridleys have been observed basking on beaches, and it is not unusual to see thousands of olive ridleys floating in front of the nesting beaches. Leatherback turtles tend to dive in a cycle that follows the daily rising and sinking of the dense layer of plankton and jellyfish.

The turtles probably feed in the upper layers of water at night. As dawn approaches, their dives become deeper as the plankton and jellyfish retreat to deeper water, away from the light of day. The turtles bask at the surface at midday when the layer sinks beyond their typical diving range. As dusk approaches, the turtles' dives become shallower as the layer rises. Green sea turtles are considered solitary, but occasionally form feeding aggregations in shallow waters abundant in seagrass or algae.

SPECIAL CHARACTERS

Sea turtles are strong swimmers. The cruising speed for green sea turtles is about 1.5 to 2.3 kph (0.9-1.4 mph). Leatherbacks have been recorded at speeds of 1.5 to 9.3 kph (0.9-5.8 mph). Forelimbs are modified into long, paddle-like flippers for swimming. Neck and limbs are nonretractile. The shell adaptations necessary for retractile limbs would impede rapid swimming. Sea turtles are excellent divers. Leatherbacks routinely dive more than 305 m (1,000 ft.). They may reach depths of more than 1,190 m (3,900 ft.) seeking jellyfish for prey. Since they are cold-blooded, sea turtles have a slow metabolic rate. This slowed metabolism allows them to stay submerged for long periods of time. Hawksbill turtles have been known to remain submerged for 35 to 45 minutes. Green sea turtles can stay under water for as long as five hours. Their heart rate slows to conserve oxygen: nine minutes may elapse between heartbeats. In the north-central Gulf of California, black sea turtles return each year to specific areas. They bury themselves in sand or mud under water and may remain dormant from November to March. During long dives, blood is shunted away from tissues tolerant of low oxygen levels toward the heart, brain, and central nervous system. Leatherbacks have high concentrations of red blood cells; therefore, their blood retains more oxygen. The muscle of leatherbacks has a high content of the oxygen-binding protein myoglobin. Myoglobin transports and stores oxygen in muscle tissue. In studies conducted on green sea turtles, lung capacity exchange in one breath exceeded 50%. Sea turtles can live in seawater with no need for a freshwater source. They obtain sufficient water from their diet and from metabolizing seawater. Like other marine reptiles and seabirds, sea turtles have a salt gland to rid their bodies of excess salt. This gland empties into the sea turtles' eyes. The secretion of salt and fluid makes them look as if they are "crying" when they come ashore. These "tears" also help keep the eyes free of sand while females dig their nests. For the most part, the only time sea turtles leave the sea is when females haul out to lay eggs. On some uninhabited or sparsely-inhabited beaches, turtles have been observed basking on land. Many adaptations that make sea turtles successful in the sea make them slow and vulnerable on land.

All reptiles, including sea turtles, have a single bone in the middle ear that conducts vibrations to the inner ear. Researchers have found that sea turtles respond to **low frequency sounds and vibrations**. Sea turtles can **see well under water** but are shortsighted in the air. Under experimental conditions, loggerhead and green sea turtle hatchlings exhibited a preference for near-ultraviolet,

violet, and blue-green light. A sea turtle is **sensitive to touch** on the soft parts of its flippers and on its shell. Little is known about a sea turtle's sense of taste. Most researchers believe that sea turtles have an acute sense of smell in the water. Experiments show that hatchlings react to the scent of shrimp. This adaptation helps sea turtles to locate food in murky water. A sea turtle opens its mouth slightly and draws in water through the nose. It then immediately empties the water out again through the mouth. Pulsating movements of the throat are thought to be associated with smelling.

REPRODUCTION

Estimates of sexual maturity in sea turtles vary not only among species, but also among different populations of the same species. Maturity may range from as early as three years in hawksbills to 12-30 years in loggerheads to 20-50 years in green sea turtles. Sexual maturity often is related to carapace size. Studies have shown that hawksbills reach sexual maturity at a carapace size of 60 to 95 cm (24-37 in.); loggerheads reach maturity at a carapace size of 79 cm (31 in.); and green sea turtles reach maturity at 69 to 79 cm (27-31 in.). Evidence suggests that some turtles continue to grow after reaching sexual maturity, while some stop growing after reaching maturity. For most species, courtship activity usually occurs several weeks before the nesting season. Two or more males may court a single female. Male sea turtles of all species except leatherbacks have enlarged claws on their front flippers. These help grasp the shells of the females during mating. Fertilization is internal. Copulation takes place in the water, just offshore.

Nesting Behaviour

Like other turtles, sea turtles lay eggs. Females come ashore on a sandy beach to nest a few weeks after mating.

- Females usually nest during the warmest months of the year. The exception is the leatherback turtle, which nests in fall and winter.
- Females of most species usually come ashore at night, alone, most often during high tide. A female sea turtle crawls above the high tide line and, using her front flippers, digs out a "body pit". Then using her hind flippers, she digs an egg cavity. The depth of the cavity is determined by the length of the stretched hind flipper.
- A female deposits 50 to 200 (depending on the species) Ping Pong ball shaped-eggs into the egg cavity. The eggs are soft-shelled, and are papery to leathery in texture. They do not break when they fall into the egg cavity. The eggs are surrounded by a thick, clear mucus.

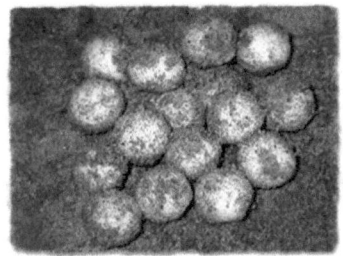

Sea turtle eggs are soft-shelled and papery to leathery in texture. A female may deposit as many as 50 to 200 (depending upon species) Ping Pong ball-shaped eggs into the egg cavity that she digs in the sand.

- The female covers the nest with sand using her hind flippers. Burying the eggs serves three purposes: it helps protect the eggs from surface predators; it helps keep the soft, porous shells moist, thus protecting them from drying out; and it helps the eggs maintain proper temperature. Experts can identify the species of turtle by the type of mound left by the nesting female and by her flipper tracks in the sand.
- Females may spend two or more hours out of the water during the entire nesting process.
- It is possible that through the storage of sperm from one or several males in the oviducts of the females, all clutches of the current nesting season may be fertilized without repeated mating.

Female Kemp's ridley and olive ridley sea turtles form masses called arribadas (Spanish for "arrival"). Arribadas contain thousands of egg-bearing females that come ashore at the same time to lay eggs.

Most females return to the same beach where they hatched to nest each year.

Recent studies suggest that some females of some species will visit more than one nesting beach (other than the original beach) in a season.

Females usually lay between one and nine clutches (groups) of eggs per season.

Females may nest every two or three years.

INCUBATION AND HATCHING

Incubation time varies with species, clutch size, and temperature and humidity in the nest. The incubation time for most species is 45 to 70 days. Research indicates that the sex of an embryo is determined sometime after fertilization, as the embryo develops, and may be temperature dependent. Lower nest temperatures produce more males; higher temperatures produce more females. Sea turtles hatch throughout the year but mostly in summer. Hatchlings use a *carbuncle* (temporary egg tooth) to help break open the shell. After hatching, the young turtles may take three to seven days to dig their way to the surface. Hatchlings usually wait until night to emerge from the nest. Emerging at night reduces exposure to daytime predators. They leave the nest and head to the water in groups. Studies have shown that some nests will produce hatchlings on more than one night.

There are several theories as to how hatchlings find the sea.

- Hatchlings may distinguish light intensities and head for the greater light intensity of the open horizon.
- During the crawl to the sea, the hatchling may set an internal magnetic compass, which it uses for navigation away from the beach. When a hatchling reaches the surf, it dives into a wave and rides the undertow out to sea.
- A "swim frenzy" of continuous swimming takes place for about 24 to 48 hours after the hatchling enters the water.
- This frantic activity gets the young turtle into deeper water, where it is less vulnerable to predators.
- There have been reports of swimming hatchlings diving straight down when birds and even airplanes appear overhead. This diving behavior may be a behavioral adaptation for avoiding predation by birds.

The First Year

During the first year, many species of sea turtles are rarely seen. This first year is known as the "lost year".

Researchers generally agree that most hatchlings spend their first few years living an oceanic existence before appearing in coastal areas. Although the migratory patterns of the young turtles during the first year has long been a puzzle, most researchers believe that they ride prevailing surface currents, situating themselves in floating seaweed where they can find food. Research suggests that flatback hatchlings do not go through an oceanic phase. Evidence shows that the young turtles remain inshore following the initial swim frenzy. Most remain within 15 km (9.3 miles) of land.

COMMENSALISM WITH BARNACLES

Sea Turtles are believed to have a commensal relationship with some barnacles, in which the barnacles benefit from growing on turtles without harming them. Barnacles are small, hard shelled crustaceans found attached to multiple different substrates below or just above the ocean. The adult barnacle is a sessile organism, however in its larval stage it is planktonic and can move about the water column. The larval stage chooses where to settle and ultimately the habitat for its full adult life, which is typically between 5 to 10 years. A favorite settlement for barnacle larvae is the shell or skin around the neck of sea turtles. The larvae glue themselves to the chosen spot, a thin layer of flesh is wrapped around them and a shell is secreted. Many species of barnacles can settle on any substrate, however some species of barnacles have an obligatory commensal relationship with specific animals, which makes finding a suitable location harder. Around 29 species of "turtle barnacles" have been recorded. However it is not solely on sea turtles that barnacles can be found; other organisms also serve as barnacle's settlements. These organisms include mollusks, whales, decapod crustaceans, manatees and several other groups related to these species. Sea turtle shells are an ideal habitat for adult barnacles for three reasons. Turtles tend to live long lives, around 50 years, so barnacles do not have to worry about host death. Secondly, barnacles are suspension feeders. Sea turtles spend most of their lives swimming and following ocean currents and as water runs along the back of the turtle's shell it passes over the barnacles, providing an almost constant water flow and influx of food particles. Lastly, the long distances and inter ocean travel these sea turtles swim throughout their lifetime, offers the perfect mechanism for dispersal of barnacle larvae. Allowing the barnacle species to distribute themselves throughout global waters is a high fitness advantage of this commensalism.

There are a few speculations however at the idea that this relationship is truly commensal. The barnacles are not parasitic to their hosts but have been found to have negative effects to the turtles on which they choose to reside. These effects however seem to depend on the quantity of barnacles affixed to its back. The barnacles add extra weight to the sea turtle, potentially increasing the energy it needs for swimming and affecting its ability to capture prey.

References

1. "National Animal". Government of India Official website.
2. "WWF - Marine Turtles". Species Factsheets. World Wide Fund for Nature. 4 May 2007.
3. Bagla, P. (1998). Indian tiger isn't 100 per cent "swadeshi". The Indian Express.
4. Biswas, S. & Sankar, K. (2002). "Prey abundance and food habit of tigers (Panthera tigris tigris) in Pench National Park, Madhya Pradesh, India". *Journal of Zoology* **256** (3): 411–420.
5. Burger, B. V.; Viviers, M. Z.; Bekker, J. P. I.; Roux, M.; Fish, N.; Fourie, W. B.; Weibchen, G. (2008). "Chemical Characterization of Territorial Marking Fluid of Male Bengal Tiger, Panthera tigris". *Journal of Chemical Ecology* **34** (5): 659–671.
6. Burton, R. (2002). *International Wildlife Encyclopedia (3 ed.)*. Marshall Cavendish. pp. 936–938.
7. Bustard, H.R. (1983). "Movement of wild Gharial, Gavialis gangeticus (Gmelin) in the River Mahanadi, Orissa (India)". *British Journal of Herpetology* **6**: 287–291.
8. Choudhury, A. (2002). *Distribution and conservation of the Gaur Bos gaurus in the Indian Subcontinent*. Mammal Review 32: 199–226.
9. Choudhury, B. C., Singh, L. A. K., Rao, R. J., Basu, D., Sharma, R. K., Hussain, S. A., Andrews, H. V., Whitaker, N., Whitaker, R., Lenin, J., Maskey, T., Cadi, A., Rashid, S. M. A., Choudhury, A. A., Dahal, B., Win Ko Ko, U., Thorbjarnarson, J., Ross, J. P. (2007). "Gavialis gangeticus". IUCN Red List of Threatened Species. International Union for Conservation of Nature.
10. Chundawat, R. S., Khan, J. A., Mallon, D. P. (2011). "Panthera tigris tigris". IUCN Red List of Threatened Species. International Union for Conservation of Nature.
11. Chundawat, R.S., Habib, B., Karanth, U., Kawanishi, K., Ahmad Khan, J., Lynam, T., Miquelle, D., Nyhus, P., Sunarto, S., Tilson, R., Wang, S. (2011). "Panthera tigris". IUCN Red List of Threatened Species. International Union for Conservation of Nature.
12. Dinerstein, E. (2003). *The Return of the Unicorns: The Natural History and Conservation of the Greater One-Horned Rhinoceros*. New York: Columbia University Press.
13. Dinerstein, E., Loucks, C., Heydlauff, A., Wikramanayake, E., Bryja, G., Forrest, J., Ginsberg, J., Klenzendorf, S., Leimgruber, P., O'Brien, T., Sanderson, E., Seidensticker, J., Songer, M. (2006) *Setting Priorities for the Conservation and Recovery of Wild Tigers: 2005–2015*. A User's Guide. 1–50. Washington, D.C., New York, WWF, WCS, Smithsonian, and NFWF-STF.
14. Duckworth, J.W., Steinmetz, R., Timmins, R.J., Pattanavibool, A., Than Zaw, Do Tuoc, Hedges, S. (2008). "Bos gaurus". IUCN Red List of Threatened Species. International Union for Conservation of Nature.

15. Eisenberg, J.F., McKay, G.M., and Seidensticker (1990), J. *Asian Elephants*. Washington, DC: Friends of the National Zoo and National Zoological Park.
16. *Ernst, C. H.; Lovich, J.E. (2009). Turtles of the United States and Canada (2 ed.). JHU Press. p. 50.*
17. *Gad, S. D.; Shyama, S. K. (2009). "Studies on the food and feeding habits of Gaur Bos gaurus H. Smith (Mammalia: Artiodactyla: Bovidae) in two protected areas of Goa" (PDF).Journal of Threatened Taxa **1** (2): 128–130.*
18. *Gee, E. P. (1959). "Report on a survey of the rhinoceros area of Nepal". Oryx **5**: 67–76.*
19. *Gee, E. P. (1963). "Report on a brief survey of the wildlife resources of Nepal, including rhinoceros". Oryx **7** (2–3): 67–76.*
20. *Godfrey, Matthew H.; Barreto, R.; Mrosovsky, N. (December 1997). "Metabolically-Generated Heat of Developing Eggs and Its Potential Effect on Sex Ratio of Sea Turtle Hatchlings". Journal of Herpetology **31** (4): 616.*
21. *Goldman, J.E. & Stevens, V.J. (1980). "The birth and development of twin Nilgai Boselaphus tragocamelus at Washington Park Zoo, Portland". International Zoo Yearbook **20** (1): 234–240.*
22. *Gopalaswamy, Arjun M.; Royle, J. Andrew; Delampady, Mohan; Nichols, James D.; Karanth, K. Ullas; Macdonald, David W. (2012). "Density estimation in tiger populations: combining information for strong inference". Ecology **93** (7): 1741–1751.*
23. Government of India (2005) *Tiger Task Force Report.*
24. *Guggisberg, C. A. W. (2001). Wild Cats of the World. New Library Press.*
25. *Hayward, Matt W.; Kerley, Graham (2005). "Prey preferences of the lion (Panthera leo)".Journal of Zoology **267** (3): 309–22.*
26. *Hildebrand, M. (1959), Motions of the running cheetah and horse, Journal of Mammalogy 40 (4), pp. 481–495*
27. *Hussain, S. A. (2009). "Basking site and water depth selection by gharial Gavialis gangeticusGmelin 1789 (Crocodylia, Reptilia) in National Chambal Sanctuary, India and its implication for river conservation". Aquatic Conservation: Marine and Freshwater Ecosystems **19** (2): 127–133.*
28. *Jerdon, T. C. (1867). The Mammals of India: a Natural History of all the animals known to inhabit Continental India* Roorkee : Thomason College Press
29. *Jethva, B.D. & Jhala, Y.V. (2004). "Foraging ecology, economics and conservation of Indian wolves in the Bhal region of Gujarat, western India". Biological Conservation **116** (3): 351–357.*

30. Karanth, K. U. (2003). *Tiger ecology and conservation in the Indian subcontinent. Journal of the Bombay Natural History Society* 100 (2&3) 169–189.
31. *Karanth, K. U. and Nichols, J. D. (1998). "Estimation of tiger densities in India using photographic captures and recaptures" (PDF). Ecology **79** (8): 2852–2862.*
32. *Karanth, K. Ullas and Sunquist, Melvin E. (1995). "Prey Selection by Tiger, Leopard and Dhole in Tropical Forests". Journal of Animal Ecology **64** (4): 439–450.*
33. *Karanth, K.U. & Sunquist, M.E. (1992). "Population structure, density and biomass of large herbivores in the tropical forests of Nagarhole, India". Journal of Tropical Ecology **8** (1): 21–35.*
34. *Khosravifard, Sam (2013). "How Many Asiatic Cheetahs Roam across Iran?". Scientific American.*
35. *King, Anthony (2015). "Can the Iranian cheetah outrun extinction?". Irish Times.*
36. *Krishna, C.Y, Krishnaswamy, J & Kumar, N.S. (2008). "Habitat factors affecting site occupancy and relative abundance of four horned antelope". Journal of Zoology **276** (1): 63–70.*
37. *Long, Barney (2014). "Irrawaddy Dolphin". World Wildlife Fund (WWF).*
38. *Mallon, D.P. (2008). "Tetracerus quadricornis". IUCN Red List of Threatened Species. International Union for Conservation of Nature.*
39. *Mallon, D.P. (2008). Boselaphus tragocamelus. In: IUCN 2008. IUCN Red List of Threatened Species. Retrieved 29 March 2009. Database entry includes a brief justification of why this species is of least concern.*
40. *Mazák, J.H., Groves, C.P. (2006). "A taxonomic revision of the tigers (Panthera tigris)" (PDF).Mammalian Biology **71** (5): 268-287.*
41. *McMaster, A. C. (1871) Notes on Jerdon's Mammals of India. Higginbothams, Madras. (pp. 123-124)*
42. *Medhi, A., Saha, A. K. (2014). Land Cover Change and Rhino Habitat Mapping of Kaziranga National Park, Assam. In: Singh, M. Singh, R. B., Hassan, M. I. (eds.) Climate Change and Biodiversity. Proceedings of IGU Rohtak Conference, Vol. 1, Part II. Springer Japan. Pp. 125–138.*
43. *Mills, S. (2004). Tiger. London: BBC Books. p. 89.*
44. *Morreale, S.; Ruiz, G.; Spotila,; Standora, E. (11 June 1982). "Temperature-dependent sex determination: current practices threaten conservation of sea turtles". Science **216** (4551): 1245–1247.*
45. *Nicolson, S.W. and P.L. Lutz. (1989).Salt gland function in the green sea turtle Chelonia mydas*. J. exp. Biol. 144: 171-184.

46. *Parliament of India. The Indian Wildlife (Protection) Act, 1972 (Substituted by Act 44 of 1991 ed.). Ministry of Environment and Forests.*
47. Pocock, R.I. (1939). "*Panthera tigris*". In *The Fauna of British India, Including Ceylon and Burma*. Mammalia: Volume 1. Taylor and Francis, Ltd., London. pp. 197–210.
48. *Ramesh, T.; Snehalatha, V.; Sankar, K. and Qureshi, Qamar (2009). "Food habits and prey selection of tiger and leopard in Mudumalai Tiger Reserve, Tamil Nadu, India". J. Sci. Trans. Environ. Technov. 2 (3): 170–181.*
49. *Rao, R.J., Choudhury, B.C. (1990). "Sympatric distribution of Gharial Gavialis gangeticus and Mugger Crocodylus palustris in India". Journal of the Bombay Natural History Society 89: 313–314.*
50. *Reeves, R. R., Jefferson, T. A., Karczmarski, L., Laidre, K., O'Corry-Crowe, G., Rojas-Bracho, L., Secchi, E. R., Slooten, E., Smith, B. D., Wang, J. Y. & Zhou, K. (2008)."Orcaella brevirostris". IUCN Red List of Threatened Species. International Union for Conservation of Nature.*
51. *Sanderson, G. P. (1907). "XVIII, XVIV". Thirteen Years Among the Wild Beasts of India: Their Haunts and Habits from Personal Observation (6th ed.). Edinburgh: John Grant. pp. 243–265.*
52. Sarma, P. K., Talukdar, B. K., Sarma, K., Barua, M. (2009). Assessment of habitat change and threats to the greater one-horned rhino (*Rhinoceros unicornis*) in Pabitora Wildlife Sanctuary, Assam, using multi-temporal satellite data. No. 46 : 18–24.
53. Schaller, G. (1967). *The Deer and the Tiger: a study of wildlife in India*. University of Chicago Press, Chicago.
54. Schaller, G. B., Simon, N. M. (1969). *The endangered large mammals of Asia*. In: Holloway, C. W. (ed.) IUCN Eleventh Technical Meeting. Papers and Proceedings, 25–28 November 1969, New Delhi, India. Volume II. IUCN Publications new series No. 18, Morges, Switzerland. Pp. 11-23.
55. Shukla, R., Khare, P. K. (1998). *Food habits of wild ungulates and their competition with live stock in Pench Wildlife Reserve central India*. Journal of the Bombay Natural History Society 95(3): 418–421.
56. *Singh, H. S.; Gibson, L. (2011). "A conservation success story in the otherwise dire megafauna extinction crisis: The Asiatic lion (Panthera leo persica) of Gir forest". Biological Conservation 144 (5): 1753–1757.*
57. Srinivasulu, C., Srinivasulu, B. (2012). *Chapter 3: Checklist of South Asian Mammals* in: *South Asian Mammals: Their Diversity, Distribution, and Status*. Springer, New York, Heidelberg, London.

58. *Stander, PE (1992). "Cooperative hunting in lions: the role of the individual" (PDF). Behavioral Ecology and Sociobiology* **29** *(6): 445–54.*
59. *Talukdar, B. K., Emslie, R., Bist, S. S., Choudhury, A., Ellis, S., Bonal, B. S., Malakar, M. C., Talukdar, B. N. Barua, M. (2008). "Rhinoceros unicornis". IUCN Red List of Threatened Species. International Union for Conservation of Nature.*
60. *Taylor, C. R.; Rowtree, V. J. (1973), Temperature regulation and heat balance in running cheetahs: a strategy for sprinters?, American Journal of Physiology 224, pp. 848–852.*
61. *Thapar, V.; Thapar, R.; Ansari, Y.(2013). Exotic Aliens: The Lion and the Cheetah in India. Aleph Book Company, New Delhi.*
62. *The Times of India (2013). "Dolphin population rises to 152 in Chilika lake in Orissa".*
63. *Whitaker, R. and D. Basu (1983). "The Gharial (Gavialis gangeticus): A review".Journal of the Bombay Natural History Society* **79***: 531–548.*
64. *Whitney, L.P (1980), The Unforgettable Elephant. New York: Walker and Company.*
65. *Wozencraft, W.C. (2005). "Order Carnivora". In Wilson, D.E.; Reeder, D.M. Mammal Species of the World: A Taxonomic and Geographic Reference (3rd ed.). Johns Hopkins University Press. pp. 532–628.*
66. *Yadav, P. R. (2004). Vanishing And Endangered Species. Discovery Publishing House. pp. 176–178.*
67. *Yamaguchi, Nobuyuki; Cooper, Alan; Werdelin, Lars; MacDonald, David W. (2004). "Evolution of the mane and group-living in the lion (Panthera leo): a review". Journal of Zoology* **263** *(4): 329–342.*
68. *Yu, Jin Hai; Xia, Zhao Fei (2013-03-01). "Bacterial infection in an Irrawaddy dolphin (orcaella brevirostris)". Journal of Zoo and Wildlife Medicine* **44** *(1): 156–158.*

References for Pictures
1. Chapters INDIAN ELEPHANT, SEA TURTLES: https://seaworld.org/en/animal-info books.
2. Chapters TIGER, CHEETAH, LION, NILGAI, BISON, ANTELOPE, INDIAN RHINOCEROS, GHARIAL, PEA FOWL: https://en.wikipedia.org/wiki.
3. Chapter IRRAWADDY DOLPHIN: wwf.panda.org

About the Authors

Sasmita Panda, born in 1982, graduated in 2002 from Godavaris Mahavidyalaya, Banpur under Utkal University, Odisha. She completed her P.G. in Zoology in 2004 and her M. Phil. in 2013 from Khallikote college under Berhampur University .She has been teaching Zoology in different U.G. Colleges since 2006. Currently, she is serving as a lecturer in zoology at Jatni college, Jatni. She has published many articles in research journals of National and International repute. She has also co-authored a Fishery book entitled "Employment Through Aquaculture" as well as a book entitled "Biology of Wild Animals" meant for students of undergraduate and post-graduate level. Furthermore, she has edited a book entitled "Water for Survival" published by M/S Nandakishore Publication.

Gagan Kumar Panigrahi, born in 1989, has passed up to CBSE from D.M. School, Bhubaneswar and B.Sc., B.Ed. from RIE (NCERT) under Utkal University. He passed M.Sc. in Life science securing 2nd position from NIT Rourkela in 2012. He also qualified in the Orissa State Talent Search Examination conducted by the Govt. of Odisha and CTET in 2011.He also qualified the CSIR-UGC NET (Lectureship in LIFE SCIENCES) in December 2012, with All India Rank of 30. He has also published research and review articles in different Journals and souvenirs of National and International repute. Currently, he is a Ph.D. scholar in Myongii deemed University, South Korea. He has also co-authored a book entitled "Biology of Wild Animals" meant for students of undergraduate and post-graduate level.

Dr. S.N. Padhi, born in 1953, obtained a B.Sc. degree in 1973, P.G. in 1975 and Ph.D. in 1981 under Berhampur University. He devoted his time in teaching and research in different Govt. and Non-Govt. colleges of Odisha since 1978. He has completed many minor research projects and major research projects funded by the University Grants Commission. He has published many research articles and edited many souvenirs of State level and National-level seminars funded by U.G.C., OBA and other Agencies. He is a recipient of the Prof. Amulya Kumar Panda Royal Teacher Award 2011 instituted by Royal College of Science and Technology, Bhubaneswar. He has edited two books on "Application Of Biology for Self employment" and "Water for Survival" published by M/S Nandakishore Publication . He also co-authored a Fishery book entitled "Employment Through Aquaculture" as well as a book entitled "Biology of Wild Animals" meant for students of undergraduate and post-graduate level. He is also a recepient of the Emeritus fellowship by University Grants Commission.